Summary

This report provides guidance on the construction applications and in-service performance of the following types of reactive polymers:

- epoxies
- unsaturated polyesters
- acrylates and methacrylates
- vinyl esters
- polyurethanes
- furanes.

It is intended to promote the more confident and cost-effective use of these materials in the United Kingdom.

The report deals with products used on site and for the factory production of resin-concrete components. The technical and economic benefits of these thermosetting materials are identified and, to assist in effective use, guidelines for selection and specification are presented. Information is also provided on the nature of reactive polymers, their formulation and their application and safety requirements.

The report has been divided into the following three volumes, anticipating the different requirements of readers with widely varying knowledge of these materials and their uses:

Volume 1: The materials and their practical applications (CIRIA Publication C537)

Volume 2: Specification and use of the materials (CIRIA PR78)

Volume 3: Materials technology (CIRIA PR79).

In Volume 1, the predominant focus is on the many established applications of reactive polymers, the main aim being to motivate consideration of these products amongst a wide engineering readership. Volume 2, intended for a more focused readership, concentrates on areas closely associated with actual use. The nature, formulation and properties of these materials are discussed in greater detail in Volume 3, which is aimed mainly at a more specialised readership requiring a greater depth of knowledge of reactive polymers.

C537

London, 2000

The use of epoxy, polyester and similar reactive polymers in construction

VOLUME 1 THE MATERIALS AND THEIR PRACTICAL APPLICATIONS

Prepared under contract to CIRIA by

S A Hurley BSc PhD

CIRIA *sharing knowledge ▪ building best practice*

6 Storey's Gate, Westminster, London SW1P 3AU
TELEPHONE 020 7222 8891 FAX 020 7222 1708
EMAIL enquiries@ciria.org.uk
WEBSITE www.ciria.org.uk

The use of epoxy, polyester and similar reactive polymers in construction.
Volume 1: The materials and their practical applications

Hurley, S A

Construction Industry Research and Information Association

Publication C537 © CIRIA 2000 ISBN 0 86017 537 5

Keywords
Reactive polymers, design, specification, site use, factory use, health and safety

Reader interest	Classification	
Design, specification, construction and supervising architects and engineers. Contractors, specialist subcontractors, reactive polymer manufacturers and suppliers	AVAILABILITY	Restricted
	CONTENT	Good practice guidance
	STATUS	Committee-guided
	USER	Construction designers, specifiers, contractors

Acknowledgements

This report is the result of a research project carried out under contract to CIRIA by Dr S A Hurley of Taywood Engineering Ltd. It provides guidance on the construction applications and in-service performance of epoxy, polyester and similar reactive polymers.

The project was developed and managed by Dr B W Staynes, research manager at CIRIA, with advice and guidance from a project steering group whose support and valuable contributions are gratefully acknowledged. The steering group comprised:

Mr D J Curtis (*chairman*)	Laing Technology Group
Mr P S Godfrey	Sir William Halcrow and Partners
Mr D R Kent	Jenkins and Potter
Prof. G C Mays	Cranfield University, Royal Military College of Science
Mr N S Moss	CIBA Polymers
Mr P B Poulson	Federation for the Repair and Protection of Structures (FeRFA)
Mr J Raymond	Scott Bader Company Ltd
Mr J D N Shaw	SBD Ltd
Mr A J Smith	Degussa Ltd
Mr L J Tabor	Consultant
Mr J A Timbrell	RESDEV Ltd
Mr E Tucker	Health and Safety Executive
Dr R West	Posford Duvivier
Mr P Woodhead	Department of the Environment, Transport and the Regions.

CIRIA and the author also appreciate the assistance given by the many individuals who were consulted during the project. In particular, the following contributions are gratefully acknowledged:

- Mr R P L McAnoy (Taywood Engineering Ltd) for collaboration in the preparation of Volume 1, Section 4
- Mr L J Tabor for the preparation of Volume 2, Section 2
- Mrs K Parkinson (Health and Safety Executive) for advice on Volume 2, Section 3
- Dr H Davies (Building Research Establishment) for voluntary comment on and amendment of Volume 2, Section 4.

The project was funded by the Department of the Environment, Transport and the Regions; Ciba-Geigy; Degussa Ltd; Jenkins and Potter; RESDEV Ltd; R M Douglas Construction Ltd; Scott Bader Company Ltd; and the sponsors of CIRIA's Core Programme.

Use of the following figures is gratefully acknowledged:

- Figure 3.11, courtesy of Sho-Bond Construction Co, Ltd
- Figure 3.12, courtesy of MBT Feb, Manchester
- Figure 3.13, courtesy of Sika Ltd, Welwyn Garden City
- Figure 3.15, reproduced with the permission of BRE/CRC Ltd, Garston/London
- Figure 3.16, extract from BS 8081: 1989 reproduced with the permission of BSI under licence number 2000SK/0090. Complete standards can be obtained by post from BSI Customer Services, 389 Chiswick High Road, London W4 4AL.

Executive summary

This report has been divided into three volumes to cater for different readership requirements. The aim of Volume 1 is to provide a stimulus for using reactive polymers and the predominant focus is upon the many established applications of these materials. Volume 2 (CIRIA PR78) contributes to the maintenance of this motivation by giving guidance on specification, good practice and the avoidance of failure. Volume 3 (CIRIA PR79) is aimed at a more specialised readership, discussing the chemical nature, formulation and related properties of reactive polymer systems. It is intended for those with responsibilities for the selection of materials and the development of innovative applications.

The generic applications discussed in Section 3 of this volume were already established in CIRIA Report R69, *Effective use of epoxy and polyester resins in civil engineering structures* (Tabor, 1978).

A brief comparison of the resin-based products currently offered by typical suppliers and those available in 1978 might also suggest few advances during the intervening period. However, a more detailed consideration of reactive polymers and their applications clearly shows that this impression is incorrect.

Significant developments have taken place in product composition and specific uses, assisted by a greater understanding of how the properties of materials can be effectively matched to application conditions and service requirements. These developments have:

- instilled greater confidence amongst designers and specifiers
- exploited the versatility of reactive polymers
- assisted clients financially.

However, the extent of these benefits varies with the particular type of application, the specific user concerned and the country of use.

User confidence

Many engineers utilise reactive polymers routinely and confidently. Others may still be reluctant, possibly due to lack of familiarity or having experienced prior difficulties.

It is mainly towards the latter and those with little knowledge of the subject that this volume of the report is directed. However, for these materials to become even more firmly established, the benefits and true limitations need to be properly appreciated. If this is to be achieved, engineers will need to develop confidence in suppliers who project their products realistically and provide appropriate support. The employment of knowledge-able and responsible applicators remains essential.

A thoughtful approach to cost is also required. As discussed in Section 4, optimum performance is not necessarily equated to the most expensive option. However, low initial cost should not be the dominant factor as it may lead to inappropriate selection and an unsatisfactory result. This emphasis is often found in the UK and is, perhaps, inherent in a competitive industry functioning in a difficult economic climate.

This tendency may decline under a variety of influences, such as:

- greater client awareness and informed demands
- audited quality systems for production and site operations
- performance standards, durability assessment and the more common use of whole-life costing
- legislation such as the Construction Products Directive and the Construction (Design and Management) Regulations (see Volume 2).

Service-life and maintenance requirements are often difficult to quantify. However, it is essential that the expectations and acceptable risks in these areas are duly weighed against both performance parameters and initial materials/application costs. It is important to select from products whose limitations are known. In addition to promoting the benefits, suppliers should also identify the limitations of their products.

Materials

The reactive polymers used in construction related applications in the UK, are based predominantly on epoxies and polyesters. The (meth)acrylates are used more extensively in other countries than in the UK. However, while they are still absent from the product ranges of many suppliers, their use has increased significantly since the production of CIRIA Report R69 (Tabor, 1978), particularly for flooring and pre-casting applications and as waterproofing membranes.

Furanes and vinyl esters are still generally confined to uses where the requirements tend towards the more extreme (temperature and chemical resistance) although the latter, in the form of urethane-acrylates, have become more widely used for flooring.

With the exception of coatings/linings/membranes and some flooring compositions, polyurethanes have remained relatively unexploited, particularly in relation to the potential they offer to vary properties for specific applications.

Many of the developments in epoxy resin technology will not be immediately apparent to the end-user. However, product components have undergone continuous development which has assisted the formulator (as discussed in Volume 3). In general, this has not led to major changes in performance, but rather to improvements in some properties and greater scope for matching formulations to specific end-uses.

The main developments in polyesters – reduced monomer emission and the control of cure shrinkage – have generally benefited factory operations such as pre-casting, rather than site use. Recent work on various (meth)acrylate resin systems, however, has provided a number of significant improvements for site applications. These include reduced health hazards and greater tolerance of poor conditions, allied to a range of useful performance properties.

Applications

The repair of reinforced concrete, following steel corrosion, has continued to evolve into an area of major worldwide activity. However, the use of resin-based mortars for reinforced concrete patch repair has undoubtedly declined over the last 20 years as improved cementitious repair products, often polymer-modified, and alternative techniques have become available.

In contrast, the contribution of resin adhesives to segmental construction is unchallenged and skid-resistant road surfacing, bridge-deck waterproofing membranes and externally bonded reinforcement are firmly established in the UK as routine applications of resin-based products. The latter technique is, however, more widely employed in some overseas countries.

Resin injection and resin-based fixing/anchoring applications have also continued to provide significant benefits and, thus, have been widely adopted. As in earlier years, surface protection, flooring and grouting are the most common applications of resin-based products.

A number of specific techniques have evolved and become accepted in the UK since the late 1970s, for example:

- coating of steel reinforcing bars with fusion bonded epoxies
- in-situ renovation of potable water pipes using spray applied epoxy linings
- in-situ sewer repair systems
- the fixing of rail tracks using tough, resilient resin grouts
- timber repair
- anti-static and conductive flooring.

Other variations on the established uses of resin-based products have been implemented overseas (see Section 3). Additional detail can be obtained from the proceedings of the various ACI and ICPIC-sponsored conferences referenced in the bibliography.

Two major trends have taken place overseas:

- a marked decline of interest in polymer impregnated concrete, due mainly to economic factors
- a significant growth in the use of pre-cast polymer concrete.

The proceedings of the more recent ICPIC conferences also suggest that environmental issues may stimulate further growth areas involving:

- resin systems derived from or incorporating waste products
- the encapsulation of hazardous waste materials
- the development of energy-saving and ecologically safer materials.

Health and safety

Health and safety is an important area of concern where reactive polymer-based products are used. Increasingly, European and UK legislation is becoming both more specific and more demanding. This issue is discussed in detail in Volume 2 and can be summarised as follows:

- familiarity with the legislation and product health and safety information must be acquired prior to commencing work
- specialist advice should be sought in any areas of uncertainty.

Increased exploitation of resin-based products

Although resin-based products have generally been accepted and used successfully in the UK, they are less exploited here than in some other countries. This project does not suggest any imbalance in technical capabilities. However, a number of relevant issues and recommendations were identified during industrial consultation. These are discussed in Section 1.4.

Provided reactive polymer-based materials are properly selected according to the application, and correctly specified and applied, they provide a range of versatile, high performance construction materials that can be used to technical and economic advantage.

Contents

List of figures

List of tables

Abbreviations

ACI	American Concrete Institute
CDM	Construction (Design and Management) Regulations
CEN	European Committee for Standardisation
CENELEC	European Committee for Electrotechnical Standardisation
CHIP	The Chemicals (Hazard Information and Packaging for Supply) Regulations
COSHH	The Control of Substances Hazardous to Health Regulations
CPD	The Construction Products Directive (89/106/EEC)
ECISS	European Committee for Iron and Steel Standards
EOTA	European Organisation for Technical Approvals; a collective organisation comprising Approved Bodies from each Member State of the EC
ETA	European Technical Approval; methods for demonstrating conformity of products to the essential requirements of the CPD where there are no existing or anticipated Harmonised European Standards
FeRFA	The Federation of Resin Flooring Formulators and Applicators; a trade association, formerly (a) the Federation of Resin Formulators and Applicators and (b) the Federation for the Repair and Protection of Structures
FRP	Fibre-reinforced plastic(s). See GRP in Glossary
GRP	Glass-(fibre) reinforced plastic(s). See Glossary
HDT	Heat distortion temperature. See Glossary
ICPIC	International Congress on Polymers in Concrete
MOAT	Method of Assessment and Testing, published by the British Board of Agrément
OPC	Ordinary Portland cement
PPE	Personal protective equipment
PAE	Polyacrylic esters
PVA	Polyvinyl acetate

PVC	Polyvinyl chloride
SA	Styrene-acrylic(s)
SBR	Styrene-butadiene rubber
SCOSS	The Standing Committee on Structural Safety; an independent body established by the Institutions of Civil and Structural Engineers to maintain a continuing review of building and civil engineering matters affecting the safety of structures
Tg	The glass transition temperature/point. See Glossary
TRADA	Timber Research and Development Association
UK CARES	UK Certification Authority for Reinforcing Steel
WIMLAS	Wimpey Laboratory Assessment Service

1 Introduction

1.1 PURPOSE

1.1.1 Objectives

This report provides guidance on the construction applications and in-service performance of epoxy, polyester and similar reactive polymer-based products, and identifies the technical and economic benefits of these materials. To assist in effective use, guidelines for selection and specification are given, together with information on the nature of reactive polymers, their formulation and their application and safety requirements.

The report has been divided into three volumes, anticipating the different requirements of readers with a widely varying knowledge of these materials and their uses. A description of the division adopted and the aim of each volume is provided in Section 1.2.

1.1.2 Background

The NEDO Report (Research Strategy Committee, 1985) highlighted the importance of research and development anticipating the future work of the construction industry and saw the development and use of new materials as important for competitiveness and efficiency in this industry. An earlier report from the Task Force (CIRIA, 1981) also identified the need for more research into materials.

Reactive polymers such as epoxy, unsaturated polyester, (meth)acrylate, polyurethane, vinyl ester and furane resins form a group of materials that are being increasingly used in the construction industry. These synthetic polymers benefit a particularly wide range of applications as they can be used on site in many forms, ranging from low viscosity solvented systems, for void impregnation, to heavily filled products not dissimilar to conventional concrete. They are additionally attractive in providing both inherently high performance and considerable scope for controlling characteristics such as the rate of cure, mechanical and adhesive properties, appearance and resistance to aggressive environments. In general, resin-based products have been accepted and used successfully (and innovatively) in the UK. However, they are being even more widely exploited in some other countries, notably Japan. A number of issues and recommendations connected with this imbalance are discussed in Section 1.4.

An earlier CIRIA report, *Effective use of epoxy and polyester resins in civil engineering structures* (R69; Tabor, 1978), contains valuable information and advice which is still relevant. However, many changes have taken place since R69 was published. The current report provides independent information and guidance to promote the more confident and cost-effective use of these materials in the UK.

1.2 DIVISION AND USE OF THE REPORT

Anticipating a wide variety of reader interest and background knowledge, the report has been divided into three volumes as follows:

Volume 1: The materials and their practical applications (CIRIA C537, this volume)

Volume 2: Specification and use of the materials (CIRIA PR78)

Volume 3: Materials technology (CIRIA PR79).

The following sections are common to each of these volumes:

- Abbreviations
- Introduction (Section 1 in each case)
- Glossary.

In addition, a complete listing of the main contents of the report is appended to each volume.

Volume 1 provides a description of reactive polymers and their general benefits and discusses some of the significant requirements in use arising directly from their chemical nature. These include the proportioning and mixing of components, and the effects of both temperature and secondary materials on the cure reactions. Key differences between reactive polymers and more traditional construction materials are also discussed, particularly with regard to stress-strain properties and the response to temperature. However, the predominant focus is on the many established applications of reactive polymers, the main aim being to motivate a wide engineering readership to actively consider the application of these products. Economic factors are also discussed in this volume.

The following applications are reviewed:

- surface treatments
- skid resistant wearing surfaces and protective overlays
- resin injection and polymer impregnated concrete
- segmental construction
- externally bonded reinforcement
- anchoring and fixing
- adhesive applications
- grouts and mortars for gap and void filling
- repair and maintenance
- floor toppings
- pre-casting
- sealing of joints and the formation of joint nosings
- waste encapsulation
- the modification of OPC concrete.

Volume 2 concentrates on areas more closely associated with the actual use of reactive polymers: specification, good site/factory practices, health and safety, and environmental issues. Consequently, a more focused readership has been anticipated. The difficulties which can result from incorrect specification and/or use are illustrated by short case histories. Relevant UK and EC legislation is also discussed in some detail in Volume 2.

Volume 3 discusses the nature of reactive polymers, their formulation and properties in greater detail than in Volume 1. This part of the report, therefore, is aimed mainly at a more specialised readership requiring a greater depth of knowledge of these materials, for example those with responsibilities for the selection of materials and the development of innovative applications. More specifically, background information on the chemical structure and formulation of resin products is given in Volume 3 for the following reasons:

- to assist communication between suppliers and users
- to illustrate the task performed by the formulator and to convey the origins of certain demands in application and service
- because some understanding of the nature of these materials can aid selection and contribute to successful use
- because a significant part of the literature dealing with these materials assumes some familiarity with their chemical basis (notably the proceedings of the conferences held by the American Concrete Institute since 1966 and by the International Congress on Polymers in Concrete since 1975).

Selection guidelines are given in this volume, mainly as tabulations which link generic polymer types to:

- the product forms which are commonly available
- the general characteristics which can be expected
- the benefits (or disadvantages) for specific applications.

More general factors connected with the selection process are also discussed.

The specific properties of reactive polymer-based products are reviewed in Volume 3, using the following sub-division:

- the unmixed and freshly mixed state
- the transition to the cured state
- the fully cured state – short-term properties
- the fully cured state – long-term properties.

1.3 SCOPE OF THE REPORT

This report deals only with the most common reactive polymers used on site and for the factory production of resin-concrete components; *viz* the following generic types:

- epoxies
- unsaturated polyesters
- (meth)acrylates – NB the term (meth)acrylates is used throughout to denote thermosetting resin systems derived from methacrylate and/or acrylate monomers
- vinyl esters
- polyurethanes
- furanes.

In Japan and some parts of Europe special hybrid types are occasionally used.

Note that the following materials, associated with reactive polymers due to a similar structure or closely related uses, are **not** dealt with in detail in this report but are reviewed briefly in Appendix A1 of Volume I:

- fibre-reinforced composites (FRP)
- formaldehyde-based polymers, eg phenolics and melamine products
- cellular polymers (foams)
- joint sealants
- polymer cement concrete.

An enormous number of proprietary products based on the resin types listed above are now available. Due to the versatility of these resins and the formulation options that they offer, these products can have a wide variety of properties. Consequently, extensive listings of specific data are not provided in this report; the property values which are given should be taken as indicative and for general guidance. For any given application, specific data for particular products should be obtained from suppliers.

Guidance on the selection of resin-based products is provided. However, the materials and uses under review are extremely diverse, as are the application and service conditions met in individual cases. Consequently, the advice on selection given here is based on a description of the materials and their main attributes and limitations. More prescriptive guidelines would require each application to be considered with a level of detail which is beyond the scope of this report.

The majority of the notable applications of reactive polymers involve cure on site. However, pre-cast resin-concrete is also discussed as it is closely related to some site-applied materials. Such resin-concrete may become particularly important in the future as pre-cast products are already widely used in a number of countries, and there is a trend to pre-assembly and modularisation.

Several on-site uses are also more common abroad than in the UK. These are reviewed because they are considered to be an important reflection of different approaches to the use of these materials.

For some applications, the selection of reactive polymers is dominated by their uniquely high performance properties. In other cases, however, economic analysis may contribute more significantly to selection as the performance of non-polymeric materials may be viable, at least in the short term. The factors associated with this analysis are discussed in Volume 1, Section 4.

Problems encountered with the use of reactive polymers can often be attributed to poor practices on site rather than to inappropriate selection or inadequate formulation. **It is essential to recognise that a complex chemical reaction is being carried out at the point of use and that certain demands are inherent in the materials and are not unnecessary restrictions imposed by the supplier.** Consequently, site use is discussed in detail (Volume 2) and is emphasised throughout the report.

Within the past ten years, legislation has been introduced which is having an increasing influence on the selection, specification and use of materials in construction. For the materials discussed here, there is additionally a clear connection between such legislation and the components used in the formulation of proprietary products.

Two areas of legislation are particularly important and are discussed in some detail:

- UK regulations which deal with detailed issues of health and safety (COSHH and CHIP) or which address this area on a broader front (CDM)
- legislation, following EC Directives, which is aimed at removing technical barriers to trade and creating a Single European Market.

There is now an extensive body of literature associated with most of the areas reviewed here. A selective approach has therefore been necessary in the use of references and in the bibliography provided for the reader requiring further information.

In addition to the open literature, conference proceedings and commercial literature, an important contribution to the report has been provided through consultation held with a number of experts and practitioners. Members of the steering group have been particularly helpful in this regard.

1.4 THE EXPLOITATION OF RESIN-BASED PRODUCTS

The consultation process referred to above included a significant element of structured discussion. In addition to technical matters, issues and recommendations connected with the exploitation of reactive polymers in the UK, as well as in the UK versus overseas, were identified and discussed.

1.4.1 UK versus overseas

Compared to several overseas countries, there appears in the UK to be an absence of any major co-ordinating influence for this technology, and particularly one which has a high level of involvement by engineers and other end-users. If little else, such a body could provide a long-term forum for users, suppliers, applicators and clients, thereby stimulating greater and mutually beneficial exploitation of these materials.

In the USA, this role has been played by the ACI since the early 1960s, resulting in a valuable series of published conference proceedings, frequently updated guidance documents/standard specifications and contributions to the development of national standards.

Although the more formal Japanese influences are less obvious, Professor Ohama and his co-workers in the Department of Architecture at Nihon University have undoubtedly been a major stimulus over the past 25–30 years.

On a wider front, the International Congress on Polymers in Concrete (ICPIC) has provided a focus since the initial meeting held in London in 1975. Given the concern regarding relative exploitation expressed above, it is perhaps pertinent to note that, whereas a reasonable number of UK contributions were made at the 1975 (London) and 1987 (Brighton) conferences, of the 100 papers presented at the 1995 Congress in Belgium, only two originated in the UK.

The fundamental reasons for different levels of exploitation are undoubtedly rooted in areas beyond the scope of this report, ie national economies, regulations, industrial structure, procurement procedures, education, academic influence and codes/standards. The influence of these factors in relation to the use of advanced composites in construction has been discussed (Head, 1994). **However, there are no technological reasons why the advantageous properties of these versatile materials should not be more widely exploited in the UK.**

1.4.2 UK issues

The more specific issues and recommendations associated with exploitation which were identified during this project are now summarised.

1. It was frequently stated during the consultation that, in general, there is a greater willingness to use reactive polymers in overseas countries compared to the UK. This active approach includes:

   - more emphasis on cost-effectiveness, as opposed to "first cost"
   - the acceptance that higher quality usually carries a cost premium
   - greater support for new ideas.

   Trust and credibility, developed by co-operation and good communication between users and suppliers, are essential. This requires clarity concerning both product limitations and application/service needs, allied to realistic expectations and proper use. Familiarity with the materials, on the part of users, assists this process.

2. Although numerous products are available, difficulty is often experienced in finding exactly what is required. Product summaries are found to be useful but they need to be supported by:

   - more concise engineering information which addresses design needs and product limitations (not merely test data of secondary relevance)
   - selection guidelines
   - demonstrated cost-effectiveness
   - better documented information on long-term performance (rather than selected case histories).

   In contrast, suppliers experience difficulties in responding to user needs as there is often:

   - a lack of adequate definition of critical properties and the required limits for given applications
   - inadequate discussion -leading to poor specification and selection
   - lack of appreciation that "ease of use" should not be equated to "tolerance of misuse".

   Certain difficulties are inherent:

   - regulations and potential liabilities (which can stifle innovation and the use of engineering judgements)
   - health, safety and environmental requirements (which are perceived as expensive to implement and a deterrent to use)
   - limits to the major improvements in site tolerance which can be achieved
   - the inevitability of formulation changes over 5–10 year periods (due , for example, to the withdrawal of raw materials, field experience, or health, safety and environmental demands).

   However, as material performance is often ahead of in-service requirements, there may be an incentive for collaboration between suppliers and users to jointly target and solve problems. A trade association such as FeRFA could play a valuable role in this respect.

The following areas requiring attention were identified during the consultation:

- improved (but not overly rigid) codes and standards which include performance classes
- test methods which support performance standards – with unified procedures for given properties
- the further development of site-based tests and, also, methods for assessing durability
- research which acknowledges site scales and conditions
- long-term and independent monitoring of site performance (quantitatively where possible) including forensic studies of in-service materials
- the consideration of "standard or generic" formulations in order to establish performance/property base-lines and limits
- critical operations which often have a low technical profile; for example, surface preparation.

3. It was frequently emphasised that effective progress is impeded if efforts in the areas discussed above are undermined by poor practices at the selection and application stages. These include:

   - meeting specifications with the lowest cost products (or controlling costs by using different suppliers for interacting products)
   - using non-specialist applicators
   - disregard of application limits due to contract timescales and/or costs
   - the use of materials which are beyond their shelf-life.

These and associated topics are discussed in detail in Volume 2 of this report (CIRIA, PR79). Here, it should be noted that site trials and demonstrations are important, as are method statements which should form part of contract quality plans. By these means, agreed benchmarks can be set.

2 Reactive polymer-based materials – an introduction to the materials and applications

This chapter provides a brief historical review, demonstrating that reactive polymers are now firmly established with a long history of use. The difficulties which have been met during their development continue to provide lessons which can be usefully applied to both current and emerging applications.

Some general characteristics and benefits of reactive polymers are discussed and a brief comparison with traditional construction materials is provided. The specific reactive polymers dealt with in this report are then described, and some general consequences of their reaction at the point of use are discussed.

2.1 REACTIVE POLYMERS IN CONSTRUCTION – HISTORICAL BACKGROUND

2.1.1 Introduction

One of the simplest classifications of organic polymers gives three main groups (see Figure 2.1):

- naturally occurring, biological polymers
- polymers obtained by modification of natural materials
- synthetic polymers manufactured from chemicals isolated from petroleum and, in earlier years, coal.

Naturally occurring polymer-based materials were used in construction activities some 2000 years ago when the Romans employed ox blood and milk as concrete and mortar admixtures to improve workability (Pomeroy, 1992). Glutinous rice paste and, also, the viscous water extract of elm shavings were combined with lime powder at least 1000 years ago, in China, to form mortar for masonry walls and for the construction of monolithic tombs (Yiun-Yuan, 1990). The use of straw in biblical times to manufacture non-cracking clay bricks may be cited as an even earlier use of polymer-based materials.

From about the mid-19th century, as the organic chemical industry came into existence, efforts were made to create new products with wider applications by modifying natural materials.

Vulcanisation, developed in 1844, transformed coagulated natural rubber latex into a far more useful material. In the early 1860s cellulose in the form of cotton waste was converted to a mouldable composition by Alexander Parkes, a Birmingham chemist. By 1868 this process had been improved and the viable commercial production of the first plastic substance, celluloid, was established in the USA. A key stage in the development of today's polymer industry occurred early in the present century when completely synthetic polymers were first produced from the controlled chemical reactions of small, relatively simple molecules. However, a further 30 years or so were to pass before a true understanding of the structure of these materials was achieved.

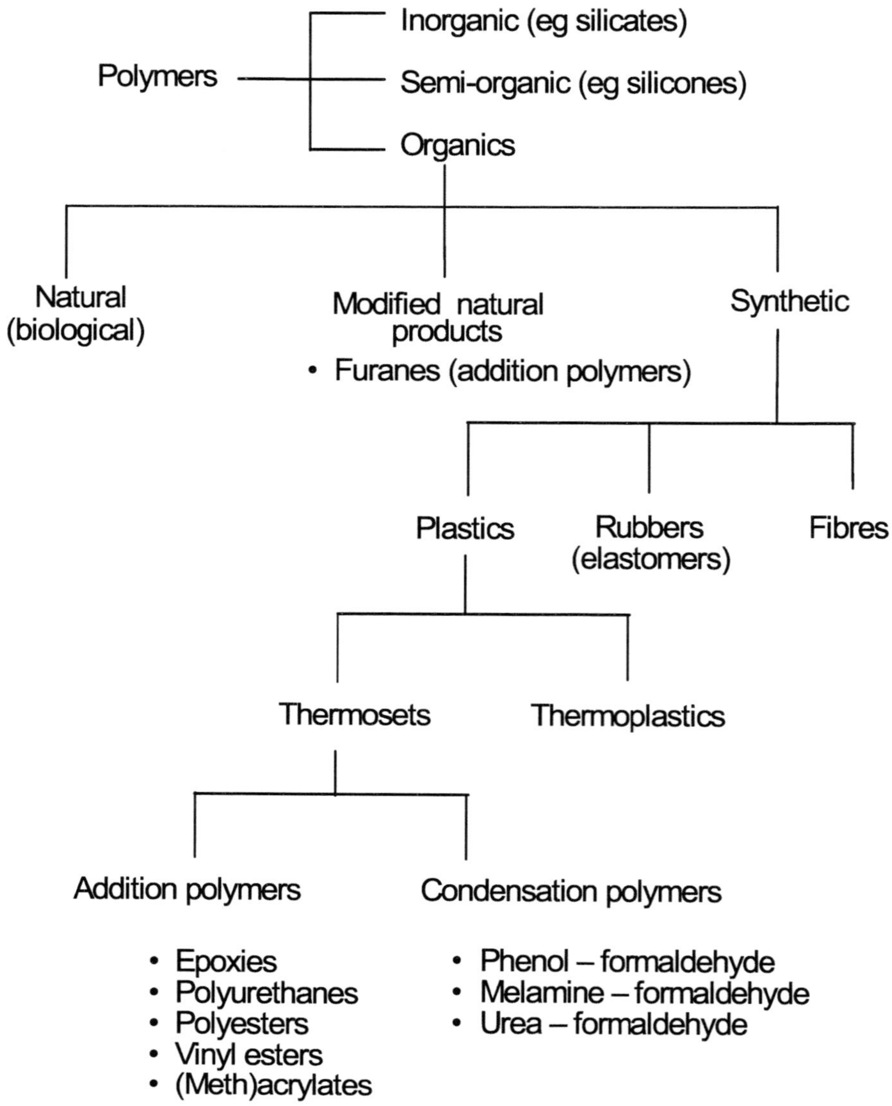

Figure 2.1 *A classification of organic polymers*

With the exception of furanes, the materials dealt with in this report are wholly synthetic, the term "resin" merely describing a viscous sticky consistency which is similar to that of naturally occurring resins.

2.1.2 Reactive polymers in construction: early days

"Bakelite", a phenol-formaldehyde based thermosetting resin, represents the first commercially available synthetic polymer. It was introduced in 1908 and used for the mass production of electrical components. Further developments led to the chemically related, and in certain respects improved, urea-formaldehyde and melamine-formaldehyde based resins in the 1930s and 1940s respectively. These polymers were used to mould all manner of domestic and industrial fittings but had little application in the construction industry, apart from secondary building fitments and laminates used for worktops and as wall linings.

In general, these resins have to be processed under factory conditions. High temperatures, often well above 100°C, are essential for the promotion of the cure reactions. In addition, the fabrication process usually requires the application of high pressures and specialist equipment to prevent the formation of voids by the small volatile molecules evolved during these reactions.

Thermosetting resins cured under normal ambient conditions and from which no species are evolved during reaction allow fabrication without the application of any significant pressure. These originated in the 1930s and 1940s. Commercial availability, which gave the opportunity for construction site applications, may be dated as follows:

- polyesters late 1940s
- epoxies mid-1940s–early 1950s
- polyurethanes 1950s
- (meth)acrylates 1960s
- furanes 1950s–1960s.

When thermosetting resins were initially used for casting under factory conditions, high additions of inert mineral fillers were used to modify physical properties and to reduce costs. Consequently, it became apparent that these materials could also be utilised on-site as binders for natural sands and aggregates, giving resin-based grouts, mortars and concretes which harden rapidly and exhibit excellent all-round physical properties, for example greatly enhanced tensile strength and chemical, impact and abrasion resistance in comparison with hydraulic cement concretes.

2.1.3 Reactive polymers in construction: developments

As material costs decreased during the 1950s and 1960s, so the range of cost-effective applications for ambiently cured resins increased. This range included both factory-based production of cast items and site-use for void-filling, surface patching, rendering, bedding and jointing.

These initial construction uses evolved from applications in other industries, as did the use of resins as adhesives and surface coatings. However, during the 1960s a number of interesting and more novel applications for reactive polymers were developed specifically for construction. Notable examples include (Potter, 1975; Mays and Hutchinson, 1992):

- thin stress-distributing weatherproof joints for post-tensioned concrete structures built from reinforced pre-cast units (segmental construction), eg Choisy-le-Roi Bridge over the River Seine near Paris (1963)
- strengthening of load-bearing concrete structures by bonding steel plates externally as additional reinforcement (first applied in South Africa and France in 1965)
- resin injection of cracked concrete structures to restore monolithic integrity and structural performance (used in California in 1959 [Boyes, 1985]) – an application which has benefited from considerable developments in metering and mixing equipment
- the provision of skid resistant road surfaces by embedding high performance aggregates (calcined bauxite chippings) in resin-based adhesive films (London 1966–1967)
- resin-bonded rock-bolts, holding-down bolts and associated fixings, and permanent ground anchors.

These and other applications of reactive polymers in construction and civil engineering are discussed in Section 3. A number of publications given in the Bibliography may also be consulted for greater detail on the early developments. In particular, it may be noted that the UK trade association, FeRFA, held a seminar in 1984 entitled "Resins in construction – 20+ years' experience".

2.1.4 Reactive polymers in construction: the present

The pioneering applications introduced in the 1950s–1960s have now become commonplace and some have been investigated in considerable detail. The range of uses for reactive polymers in construction has also continued to increase although, in some areas, resins have been replaced by improved cementitious products (often polymer-modified), eg mortars for the patch repair of reinforced concrete.

This evolution has not taken place without some difficulties. However, when a failure does occur, the problem usually has a very specific and identifiable origin, for example:

- poor workmanship
- selection of an inappropriate product
- lack of knowledge of the detailed requirements of a given application
- insufficient appreciation of the limitations of a given product or resin type (whether by the manufacturer, the specifier, or the user).

Experience and advances in materials have given increased confidence and the world-wide status of reactive polymers in the construction industry is firmly established. However, while these materials continue to be used extensively by the UK construction industry, they are being more widely exploited in some other countries, particularly as pre-cast products (Mikhailov et al, 1992; see also the various Proceedings of the International Congress on Polymers in Concrete given in the Bibliography).

The range and number of both proprietary products and their components has also increased enormously. Consequently, users and specifiers have benefited from the availability of many products which are carefully tailored to given application and service conditions. However, this has also led to problems of selection.

Information on properties and performance has inevitably increased, particularly as many case histories of a reasonable duration are now available. However, in the UK (although less so in some other countries, eg Japan/Germany) the introduction of standard test methods and specifications, and also the provision of long-term performance data, has tended to be lacking in comparison with the rate at which "new products" have become available. For example, BS 6319, *Testing of resin compositions for use in construction*, was not introduced until 1983, at least some 20–30 years after many of the products to which it applies (although test methods produced by the Trade Association FeRFA were available at an earlier date [Tabor, 1978]). The above British Standard is now available in 12 parts under the new title, *Testing of resin and polymer/cement compositions for use in construction*.

Within Europe, the approach to standards has changed dramatically in recent years following legislation directed towards the elimination of technical barriers to trade and the establishment of a Single European Market (BRE, 1995). In particular, the Construction Products Directive has led to considerable activity, by both national and European organisations responsible for standards. This is directly connected with many of the applications discussed in this report.

The exact performance requirements for a number of applications are still evolving towards a clear definition. However, the early projection and use of resin-based systems on the simple and often misleading basis of their being "at least 2–3 times stronger than concrete" has now all but disappeared as the true benefits and limitations of yesterday's advanced materials have become more concisely appreciated.

These areas are considered in greater detail in Section 3.

2.2 AN INTRODUCTION TO THE RANGE OF REACTIVE POLYMERS AND THEIR BENEFITS

2.2.1 General characteristics

As cured thermosets, unlike thermoplastics, cannot be reshaped by heat or pressure, fabrication for service is accomplished by placing the fluid or plastic resin system immediately after mixing the components supplied (commonly two or three). Depending on the application, brushing, spraying, trowelling, pumping, or pouring may be used. Conversion to a cross-linked structure and the required solid form then follows by irreversible chemical reaction.

Rather than presenting a problem, this inherent processing requirement confers versatility. It permits the thermoset to be fabricated on site under a range of ambient conditions – an invaluable characteristic in many civil engineering and construction applications where placing and fixing a preformed rigid shape would be impossible or undesirable. Consequently, resin-based systems provide unique practical and economic solutions to a wide variety of construction tasks. Such benefits are readily apparent in their use as structural adhesives or as crack filling materials – applications where few, if any, alternatives are available.

The broad generic range of reactive polymers gives scope for achieving a correspondingly wide range of characteristics, both prior to cure and in-service. Furthermore, as at least one component of a thermosetting polymer system is normally a liquid, both solid and liquid modifying materials may be readily added to give further diversity. Thus, their form prior to hardening can vary from fairly low viscosity liquids to highly filled and non-flowing compositions such as mortars and concretes. Intermediate forms, such as coatings, waterproofing membranes and thick lining materials, flowing grouts and pastes, can be readily produced. In addition, the low viscosity extreme can be extended by the addition of solvents to give systems which can impregnate porous solids such as wood, brick and concrete. This potential for modification provides the basis for proprietary products which are tailored to many specific applications and to cost requirements.

2.2.2 Variations in formulation

In contrast to materials which modify the resin system by their physical presence, others have a direct chemical effect on the cure reaction and may become part of the cross-linked structure, eg accelerators, retarders and reactive diluents.

The rate of cure can also be controlled by varying the chemical nature of the reactive resin and co-reactant (catalyst, hardener, or curing agent). Sufficient scope is usually available for a specific product to be formulated in different grades with relatively insignificant variations in service performance. Consequently, use under UK winter conditions can be accommodated, in addition to that in very hot climates (Figure 2.2).

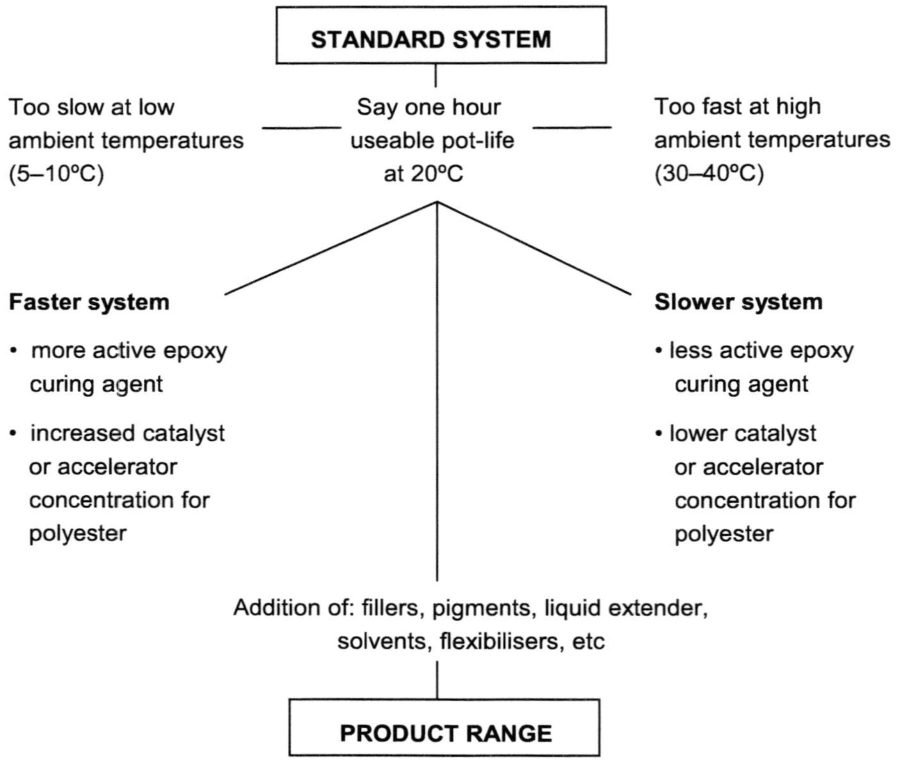

Note: this scheme is deliberately simplified to convey some basic features of formulation. It is not intended to imply that modifications can easily be carried out by an end-user (or by the formulator).

Figure 2.2 *Derivation of a product range*

The basic elements in a thermosetting resin system, ie the reactive polymers, may be viewed as providing a structural skeleton or framework. By analogy with the cladding and detailing of a building, the more specific characteristics required for given applications are then obtained from additional materials. The present availability of reactive polymers and co-reactants is such that, for most generic resin types, a virtually unlimited range of chemical structures can be produced. Consequently, the service properties can be varied very widely, although there are certain practical constraints.

It must be emphasised that this diversity does not give the user a correspondingly wide latitude to interchange generic types, or specific products, as not all marketed resin systems are suited to a given application.

Proprietary resin-based products are most commonly generated by formulators who act as a link between the end-user and the chemical companies who manufacture the resins, curing agents, catalysts, and additives. The user has little scope to produce or modify resin systems to suit his particular needs. Consequently, he must select a proprietary product from those formulated for a given application. This supply chain, which also exists with other site-applied products, such as joint sealants and concrete admixtures, is viable if all parties accept certain responsibilities.

In general, the end-user does not possess sufficient specialised knowledge and experience, or control facilities, to deal with the detailed chemistry involved in the selection of components and the formulation of ready to use products. The responsibility for these tasks usually lies with formulators who are thus not merely sub-dividers of bulk raw materials. They have a key influence on the development and confident use of these products, and must select components with appropriate judgement to provide complete and balanced compositions.

The formulator must also provide products which:

- can be satisfactorily handled and cured within recommended limits, especially those of temperature and exposure to moisture
- have suitable properties for the service conditions envisaged
- minimise health risks and conform to Health and Safety Regulations
- are supported by appropriate data and quality assurance.

Once selected, a resin system must be used according to the formulator's guidelines. All reputable manufacturers provide comprehensive instructions for the use of their products (including information on health and safety) and these should be carefully read and understood. The applicator must appreciate that a chemical reaction is being carried out at the point of use, and that all such reactions are affected by the surrounding conditions, particularly temperature. Thus, use at an unacceptably low temperature may lead to a very slow rate of hardening and failure to attain the predicted service properties; use at an unduly high temperature may give insufficient time to place the material satisfactorily prior to hardening. Such temperature limits will vary with the formulation as well as the resin type.

Further discussion concerning both the formulation and use of reactive polymer systems is provided in the following sections/volumes of this report. Individual properties are discussed in some detail in Volume 3. However, it is useful at this point to summarise the broad features of these products and, also, to give a brief comparison with several more traditional construction materials (see Tables 2.1 and 2.2).

Table 2.1 *Relative advantages and disadvantages of typical resin-based systems*

Resin type	Benefits	Disadvantages *
Epoxies	Excellent adhesion, good chemical resistance, negligible post-gelation cure shrinkage, relatively tolerant of cold/wet cure conditions, good mechanical properties, wide scope in formulation, wide variation in pot-life and cure time, good shelf-life.	High viscosity, health and safety, chalking caused by sunlight, slow cure, limited high-temperature performance, high thermal coefficient of expansion, accurate proportioning and complete mixing critical.
Poly-urethanes	Good chemical resistance, wide range of flexibility, good mechanical properties, good adhesion, wide formulation scope.	Health and safety, water sensitivity, high thermal coefficient of expansion.
Polyesters	Fast cure, low-temperature cure, mixing tolerance, chemical resistance good against acids, good mechanical properties, low viscosity.	High cure shrinkage, adhesion limited by shrinkage, health and safety, shelf-life can be short, chemical resistance against alkalis can be poor, high exotherm, high thermal coefficient of expansion.
Vinyl esters	Similar to polyesters but with improved chemical resistance and elevated-temperature performance.	Similar to polyesters but with improved chemical resistance.
Acrylates and methacrylates	Similar to polyesters but cure shrinkage has less consequence, good intercoat/interlayer adhesion, good alkali resistance, very low viscosity.	Similar to polyesters, relatively poor solvent resistance.
Furanes	Outstanding acid resistance and high temperature performance.	Sensitive to alkaline conditions/ surfaces during cure, brittle, high coefficient of thermal expansion.

* The health and safety requirements of all resin-based systems need to be considered (see Volume 2).

Table 2.2 *Resin-based systems and traditional construction materials: comparison of typical properties*

Property (at 20°C)	Unit	Epoxy system	Polyester system	OPC system	Mild steel
Relative density	–	0.9–2.0	1.2–2.0	2.2	7.8
Unconfined compressive strength	MNm^{-2}	55–150	55–150	15–70	400
Flexural strength	MNm^{-2}	25–50	25–30	2–5	–
Tensile strength	MNm^{-2}	10–25	10–25	1–5	400
Young's modulus	GNm^{-2}	0.5–30	1–20	20–40	210
Poisson's ratio	–	0.4	0.4	0.2	0.3
Tensile elongation at break	%	0–15	0–6	0.1–0.2	30
Impact resistance	Nm	30	–	16	27
Approx. work to fracture	Jm^{-2}	100	–	20	10^5–10^6
Linear coefficient of thermal expansion	10^{-6}/°C	25–35 (100 unfilled)	25–35 (100 unfilled)	7–12	11
Glass transition temperature	°C	30–70 (100+ if post-cured)	30–70 (100+ if post-cured)	–	–
Effect of elevated temperature on mechanical properties	–	Often a marked loss of strength and decrease in modulus at 40–80°C	As epoxy	Little at <300°C	Little at <200°C
Effect of low temperature on mechanical properties	–	Often a marked increase in modulus and decrease in ultimate elongation	As epoxy	Of no concern	Brittle fracture
Creep	Micro-strain	200–10 000+ (depending on stress/proximity to Tg)	As epoxy	500–1000	–
Chemical resistance:					
Solvents		Excellent to poor	As epoxy	Excellent	Excellent
Acids		Excellent to good		Very poor	Poor
Alkalies		Excellent to good		Excellent	Excellent
Fire resistance		Relatively poor	As epoxy	Good	Good
Water absorption (28 days)	% w/w	0.2–5%	As epoxy	5–15	0
Rate of useable strength development	Days or weeks	1–7 days	<1 day	1–4 weeks	–
Effect of low temperature on cure	–	Problems at <5 °C Occasionally <10 °C	May be cured down to about -20°C	Problems at <5°C avoided by well-established practices	–
Intrinsic adhesive capability	–	Excellent	Very good	Poor	–
Cure shrinkage	–	Very low to negligible	Significant	Significant	–
Colour/surface texture	–	Wide variation possible	As epoxy	Some variations possible	Little
Relative cost	–	High (x5–100)	As epoxy	Low (xl)	–

Notes

1. Resin-based systems include both unfilled and filled compositions.
2. For resin systems in particular, very wide ranges in properties can be achieved.

2.2.3 Terminology

Although a glossary of terms is appended, several in common use are noted below. Their applicability to resin-based systems is indicated, although they are also used for other polymer types (see below).

Polymer-concrete, PC	A blend of reactive polymer binder and fillers or resin impregnation of pre-packed aggregate.
Polymer-impregnated concrete, PIC	Resin impregnation of Portland cement concrete.
Polymer (Portland) cement concrete, P(P)CC, or polymer-modified concrete, PMC	Modification of an aqueous slurry of (Portland) cement concrete with a resin system.

In the case of polymer-concrete, aggregate may be pre-packed in either a mould or a construction void. The term P(P)CC is most frequently used to describe modification by an aqueous dispersion of a thermoplastic polymer or synthetic rubber (see Appendix A1.5). However certain types of polyesters and water-based epoxies may also be used.

2.3 REACTIVE POLYMERS AND TRADITIONAL CONSTRUCTION MATERIALS – A BRIEF COMPARISON

2.3.1 Key differences

Reactive polymers may replace, or perhaps more commonly supplement, the more traditional materials of construction such as concrete, steel, wood and masonry. In order for the architect, designer or engineer to make effective use of such materials, knowledge of their performance under a variety of environmental and loading conditions is required. Some appreciation of the relationship between test data and service performance is also useful.

For the more traditional materials, familiarity and proven use over a very long period is established. The relative lack of data and experience for reactive polymers can place a greater emphasis on a basic understanding of the materials and their properties in order to achieve effective use. Reactive polymers have many markedly different properties to traditional materials (see, for example, Figure 2.3) and, when used in combination, the potential consequences of this mismatch need to be considered. In certain applications, for example protective coatings, differences in properties are of course essential.

Although expressed some 30 years ago, the thoughts of Professor E. H. Andrews given in the introduction to his book *Fracture in polymers* (1968) still have relevance to the use of polymers in construction:

"The physics of polymers is a neglected field and this is undoubtedly due to the lag of science behind technology which so often occurs when discoveries or inventions (in this case those of the polymer chemist) are promptly and successfully exploited. The technological 'explosion' in the field of polymers has been truly impressive and some of today's commonest plastics were unknown 10 years ago. It is only now, as engineers are becoming increasingly interested in structural applications of plastics, that the gaps in our physical understanding of these materials are being exposed. It is not enough to rely on past experience with other solids such as metals, polymers exhibit a whole host of unfamiliar and often unexpected properties."

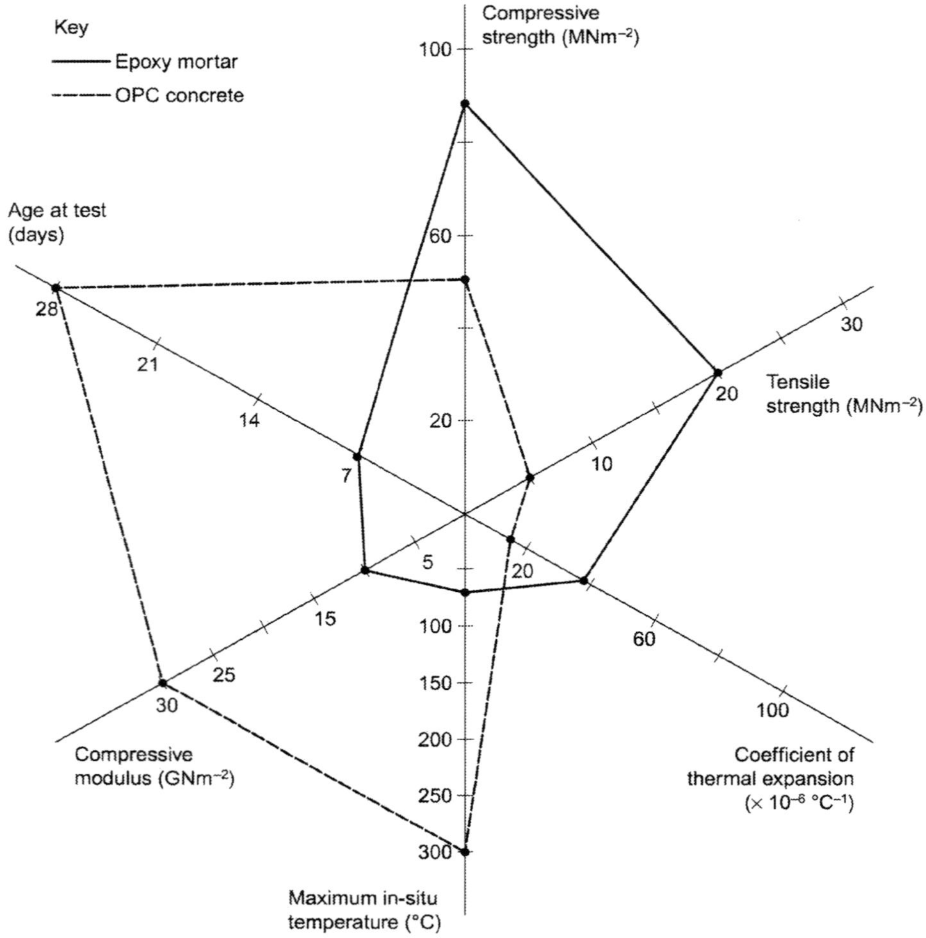

Note: 7 days and 28 days are the periods commonly used for testing because a high proportion of the ultimate properties are obtained at these times.

Figure 2.3 *Property comparisons: resin mortar and Portland cement concrete*

Thus, for example, response to stress over long periods, the effect of temperature changes and levels, fire performance and general durability in the service environment may require special consideration.

It is essential that test results are applied with some understanding of the basic nature of the materials, particularly as there is now such a proliferation of products which appear to be very similar but which are often, in fact, subtly different. For example, with regard to durability, a design life may be desired that is well in excess of the time for which the polymer has been available. Inevitably there are limitations to relatively short-term laboratory tests, simulations and field trials, which are of necessity usually restricted in scope. Consequently, the results of research can play a key part in the decision to use such a material. Nevertheless, no matter how carefully such research has been carried out, some doubts concerning long-term performance may still remain.

Steel, wood and brick possess their ultimate properties at the time of use, but reactive polymers used on site develop their properties in situ. Although the final product is mainly a consequence of the specific composition, the conditions under which the cure reaction takes place (temperature, mass, and, in some cases, humidity) can also influence performance significantly.

Compared with the traditional construction materials, the properties of most ambiently-cured resins are generally much more sensitive to relatively small temperature changes. When stressed, polymers also have a greater tendency to exhibit creep, stress relaxation and, in some cases, creep rupture.

2.3.2 Stress-strain properties

The broad spectrum of cured resin properties is illustrated by the stress-strain diagrams given in Figure 2.4. This type of variation can be observed with one specific resin as a 20–30°C change in temperature may transform a material which is stiff and strong at 20°C, to one which is stiff but more brittle at 0°C, or flexible and much weaker at 50°C. Similarly, the absorption of moisture by some cured resins results in a lower effective modulus as water can act as a plasticiser.

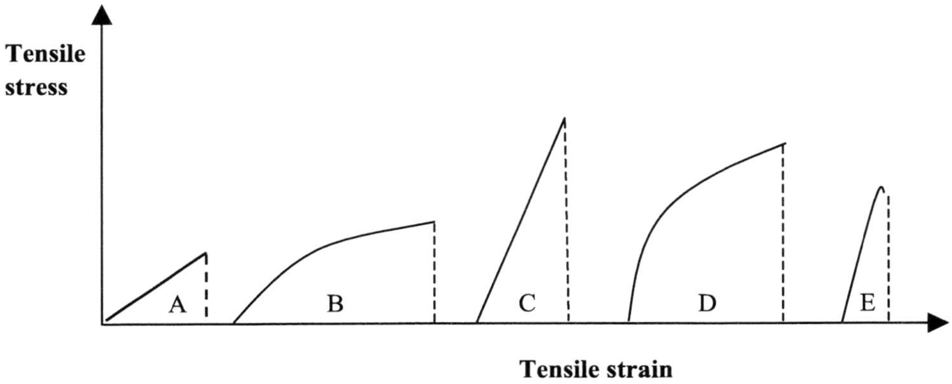

Key
A Flexible and weak
B Flexible and tough
C Stiff and strong
D Stiff and tough
E Stiff and brittle

Note: This range of properties can be achieved through variations in polymer type and/or formulation.

Figure 2.4 *Illustrative tensile stress-strain curves for several types of resin system*

Due to their molecular structure, polymers display the combined characteristics of elastic materials and viscous liquids, ie they are viscoelastic, giving a "delayed elastic response". These effects are often particularly marked at higher temperatures. Due to this time dependence, many measured mechanical properties are particularly sensitive to the rate of strain used in testing. Lower tensile strain rates, for example, will often lead to a higher ultimate elongation (but a lower ultimate stress) as the polymer has sufficient time to overcome the applied stresses by relaxation effects (Figure 2.5).

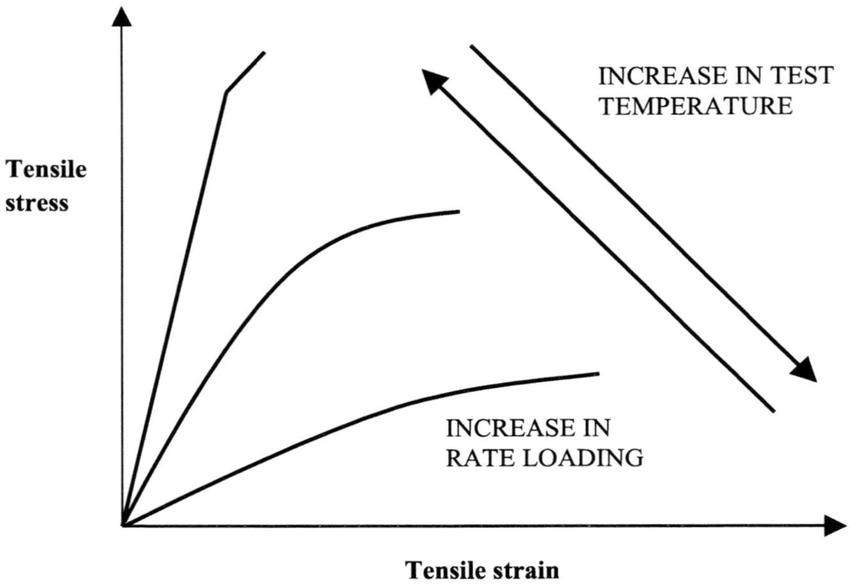

Figure 2.5 *An illustration of the response of polymers to changes in temperature and rate of loading*

In service, the consequences of viscoelasticity may be disadvantageous, or beneficial, depending on circumstances. If high levels of creep under sustained load are unacceptable, then a resin which exhibits such characteristics under the service conditions should not be used. However, this same product could be particularly suited to a different application where stress relaxation effects are desirable, eg to avoid failures due to sustained stress concentrations or dynamic loads. In a third use where high stress transfer is essential, the same material would once again become unsuitable.

2.3.3 Response to temperature

One unique property of organic polymers, arising from their chain-like (or cross-linked) structure is the glass transition temperature (Tg). Although discussed in some detail in Volume 3 of this report, it is briefly introduced here as it is one of the most important and fundamental characteristics of these materials (and some inorganic glasses), but one that is not exhibited by concrete, steel or wood. For reactive polymers, the magnitude of Tg depends to a marked extent on the specific cure conditions and this, in turn, affects the in-service performance.

When a polymeric material below its glass transition temperature is heated, it will demonstrate a fairly sharp transition from a hard solid to a softer, more flexible, rubber-like material, as illustrated in Figure 2.6.

This transition point corresponds to the onset of significant motion of quite long segments of molecules, thus characterising a rubbery state where viscoelastic effects are more pronounced. Below Tg, this motion is, in effect, frozen, and only relatively small parts of the molecules are able to move against local restraints. The consequence of cooling below Tg is demonstrated by the effect of impact on rubber cooled in liquid nitrogen: it is very brittle, or glass-like, and shatters easily. However, in the absence of impact, the glass/rubber transition can be reversed indefinitely without detriment to the material. The glass transition temperatures of some common polymers are given in Table 2.3.

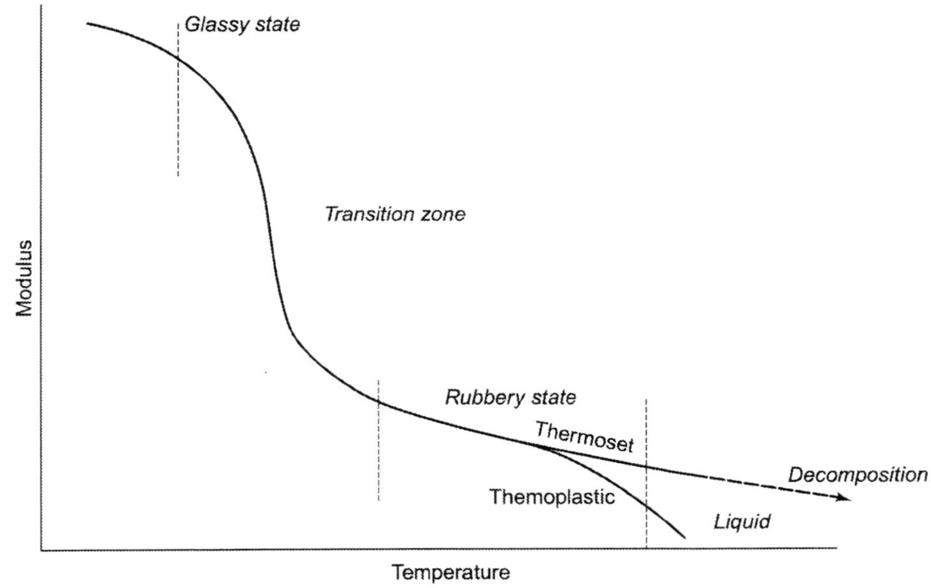

Figure 2.6 *Polymer modulus as a function of temperature*

Table 2.3 *Typical glass transition temperatures**

Polymer	**Glass transition temperature** (°C)
Silicone rubber	-123
Polyethylene	-120
Natural rubber	- 70
Nylon	50
PVC (unplasticised)	87
Hot-cured epoxy	100+ (commonly)
Polycarbonate	150
Typical reactive polymers used in civil engineering (ambient cure):	
epoxy	30–60
polyurethane	20–70
polyester	30–70+
vinyl ester	30–70
(meth)acrylate	30–60
furane	70+

* May vary with compounding. Common figures are given for reactive polymers.

Although the mechanical properties of a polymer are determined essentially by whether it is above, or below, its glass transition temperature, secondary factors also affect mechanical behaviour. Thus, some materials below Tg can be tough rather than brittle, eg nylons and polycarbonates which have high impact resistance below Tg. Additionally, although viscoelastic effects are less pronounced below Tg, creep and stress relaxation may still occur.

Around the glass transition temperature many properties change significantly, eg moduli, coefficient of expansion, strength, adhesion, creep and stress relaxation, damping or dynamic mechanical properties, and the permeability with respect to gases and vapours.

Such effects are illustrated in Figures 2.6–2.8. For thermosetting resins (in contrast to thermoplastics) liquid flow above Tg is prevented by the cross-linked nature of these systems. If a continuing increase in temperature is imposed, then both types of polymer will undergo permanent decomposition.

Two particular effects on service performance stem from the phenomenon of a transition temperature.

1. It is often necessary to limit the upper service temperature of a resin system, particularly in structural applications.
2. Where a low modulus, rubber-like flexibility is essential then the polymer selected should have a Tg below the service temperature, eg sealants, membranes.

Within limits, the formulator can adjust the specific components in a resin product in order to control the Tg, (although it is also affected by the cure conditions, as discussed in Volume 3 of this report).

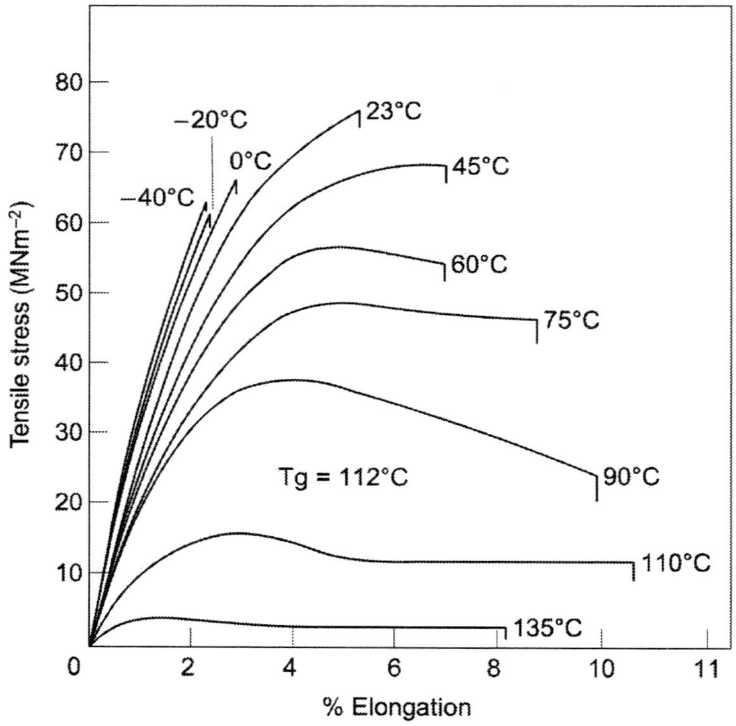

Cure conditions: 4 days at 60°C + post-cure at 190°C

Figure 2.7 *Stress-strain curves for diethylenetriamine cured epoxy resin (Tg = 112°C) at various test temperatures (Lewis and Ramsey, 1966)*

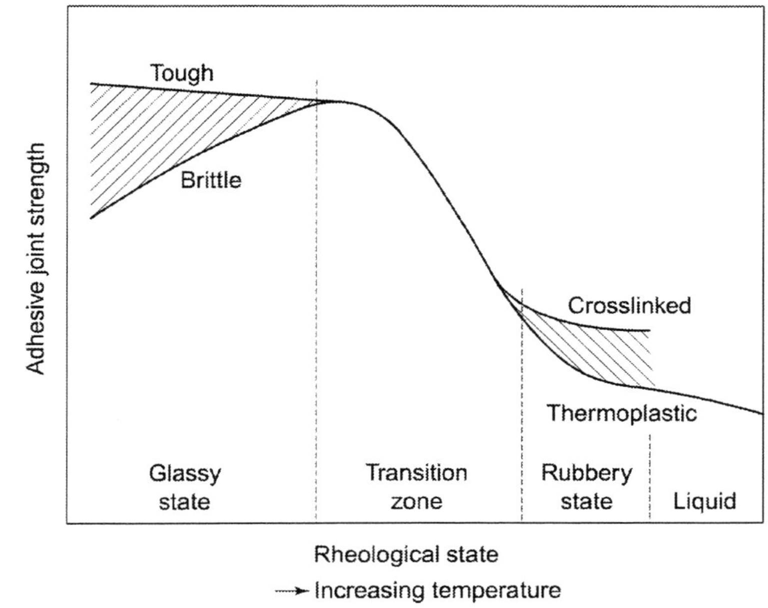

Figure 2.8 *Effect of rheological state on adhesive lap shear and tensile joint strengths (May and Tanaka, 1973)*

2.4 REACTIVE POLYMERS – GENERIC TYPES

2.4.1 Introduction

The following sections (2.4.2–2.4.5) give a brief description of the generic resin types considered in this report. Additional detail is given in Volume 3, although the account is necessarily both simplified and illustrative due to the chemical complexities and numerous variations associated with reactive polymers. For a full description, the publications given in the Bibliography (Volume 3) should be consulted.

One final introductory point, on terminology, is needed. This is potentially confusing as both the cross-linked polymers and the pre-polymers are commonly described as "resins". Thus, one component of a two-pack adhesive is an epoxy resin which, on reaction with the hardener, gives a cured material which is also referred to as an epoxy resin. Strictly, the term "resin" should be applied only to the precursor – the cured material is simply an "epoxy". This dual use is widespread but the specific context usually conveys the exact meaning.

2.4.2 Epoxy resins

Epoxy resins were first synthesised in the late 1930s, with commercial manufacture by CIBA A.G. of Switzerland (now CIBA Polymers) and Shell Chemicals commencing in the early 1950s. Epoxies are one of the major thermosetting resins, with extremely diverse applications, which utilise their excellent adhesion to a wide variety of substrates, a high strength:weight ratio, outstanding toughness, high electrical resistance and excellent chemical resistance.

A number of specific resins are used in products formulated for construction uses but in all cases the reactive and generic epoxy group is situated at each end of the resin molecule. Typical epoxy resins cannot be cross-linked via these groups by heat alone – even heating to 200°C has little effect. To obtain a network structure, a co-reactant (curing agent) must be added.

The selection of this curing agent is most important as it has a critical effect on virtually every property of the product. It has been estimated that there are well in excess of 500 curing agents on the European market. Fortunately, many fall into reasonably well-defined groups with known characteristics thus easing selection (see Volume 3). For construction applications, the curing agent is invariably an organic amine.

A notable feature of the cure reaction is that low molecular weight volatile materials are not evolved, thus avoiding the formation of voids in the solid. As the amine reacts by addition to the resin, becoming a permanent part of the network, a certain resin/curing agent mix ratio is required to obtain an optimum degree of cross-linking and a good balance of properties. This ratio is recommended by the supplier and is often based on experimental evaluation as well as theoretical calculations. Misproportioning by the user, usually beyond the ± 10 per cent level, is likely to give inferior properties.

The resin and curing agent must also be thoroughly mixed so that the reactive groups are brought into close proximity. Poor mixing may allow sufficient reaction for hardening. However, the required service properties may not be obtained.

Epoxy resins are the most widely exploited generic type of reactive polymer in civil engineering and construction as their basic chemistry allows attainment of:

- virtually any form of product
- properties which:
  - are inherently suitable for many applications
  - can be widely varied
- a low risk of fire (non-solvented products)
- a long shelf-life.

Aside from their general versatility, two properties in particular have probably contributed to the widespread use of epoxies, ie:

- the lack of cure shrinkage
- the ability of suitable formulations to bond well to wet substrates (some products can be used underwater).

Although epoxies are thus the generic type initially considered (and used) for many applications, the following drawbacks may lead to the selection of alternatives:

- relatively slow cure compared to other resin types, particularly at low temperatures
- relatively low tolerance of poor proportioning/mixing – less suited to some forms of convenience packaging (see Volume 3) and pack-splitting on site
- inadequate performance in some specifically demanding service conditions (aggressive chemicals/elevated temperatures)
- relatively high viscosity (non-solvented products) compared to polyesters and (meth)acrylates in particular
- higher cost than some other types of reactive polymers.

2.4.3 Unsaturated polyesters, vinyl esters and reactive (meth)acrylates/acrylates

Although this group of thermosetting resins have significant differences in chemical structure, they are presented under a common heading as cross-linking is achieved via the same basic mechanism. The demands placed on the user by each type are therefore somewhat similar.

Unsaturated polyester resins

Polyesters are defined generally as polymers which contain the recurring ester group in the main chain. A wide variety of polyester-based materials is available commercially, ranging from surface coatings (alkyds) to fibres. Those of interest here are linear polymers containing certain reactive groups which provide sites for subsequent cross-linking (it is these groups which are referred to by the term "unsaturated"). Such polymers first became available in the USA in 1946, and by about 1949 their commercial use for glass fibre-reinforced laminates was firmly established.

The unsaturated polyester itself is a thermoplastic solid at room temperature and has limited use. The molten polymer could be cross-linked directly but reaction would be slow and the resultant structure would have poor properties. To avoid these limitations, the polyester is dissolved in a reactive solvent during manufacture. While providing a useable viscosity, this additional component, most commonly styrene, is also a cross-linking agent. The resins used to formulate proprietary products are, therefore, solutions of the reactive polyester (65–70 per cent) in a reactive monomer/solvent (30–35 per cent).

In contrast to epoxies, where the resin and curing agent are mixed immediately before use, these liquid polyesters contain all of the essential reactive groups for cross-linking. The reaction itself is induced by a triggering mechanism.

Heat, or light, may act as the trigger and, for this reason, unsaturated polyesters usually have a shorter shelf-life than epoxies, even at normal ambient temperatures. Stabilisers are added during manufacture to avoid premature self-gelation. Stability is further enhanced by storage at low ambient temperatures in containers which exclude light.

Heat, or light promoted cross-linking is difficult to control on site. Consequently, polyesters are cured by adding a low concentration (circa 1 per cent active matter) of an organic peroxide catalyst immediately before use (this catalyst is often referred to as an initiator).

When the catalyst is used alone, heat is usually necessary for an acceptably rapid cure; temperatures of about 70–120°C are required. To induce cure at normal and low ambient temperatures, a soluble accelerator (activator or promoter) is added to the resin. Reaction between the catalyst and accelerator generates species which, in turn, induce reaction between the polyester and the monomer, thus giving cross-linking.

Although the peroxide catalyst is always added immediately before use, accelerators can be incorporated during the manufacturing process while still retaining a good shelf-life. Such pre-accelerated resins have obvious convenience and are hence readily available.

Where accelerators are handled by the user, they must always be stored and added to the resin quite separately from the peroxide catalyst. Direct mixing of the two can lead to a violent explosive reaction.

The peroxide catalysts are supplied in a diluted form, rather than as the much more hazardous pure materials; for example, as pastes (diluted with a plasticiser) or powders (dispersed onto calcium sulphate). The former are convenient when the resin-based product is a liquid or lightly filled paste. However, with more highly filled two-pack grouts and mortars, the peroxides in powder form can be pre-blended with the fillers.

Polyesters, in general, are slightly more tolerant of site misuse than epoxies. Provided they are not too extreme, variations in the catalyst level tend to affect the gelation and cure time rather than the service properties. Also, as all the reactive groups are intimately pre-mixed in the liquid resin, cross-linking can proceed acceptably even when the catalyst and resin are mixed somewhat less than thoroughly.

The following characteristics of polyesters provide particularly useful benefits:

- very rapid cure rate and good cure at very low temperatures
- very low viscosity (unfilled)
- some tolerance of less than ideal proportioning and mixing (pack splitting for repetitive low volume use may be possible)
- relatively good performance at elevated temperatures (including chemically aggressive conditions – acidic in particular)
- relatively low cost.

Particular attention should be given to the useable life of polyesters. This must be more than adequate for the anticipated working conditions as transformation to the solid phase generally occurs very sharply. With epoxies and polyurethanes, there is usually a more gradual increase in viscosity between mixing and the gelation point.

Selection of a different resin type may be necessary due to the following factors:

- high cure shrinkage (in both the liquid and solid state); this may be exacerbated by thermal contraction as the peak of the reaction exotherm usually occurs after gelation. Loss of adhesion/delamination, cracking or high internal stresses can follow
- adhesion to moist substrates is often relatively poor
- long-term chemical resistance (and maintenance of adhesion) may be limited in wet alkaline conditions
- low flash point and odour/toxicity of vapour (although styrene emission can be reduced with certain additives (see Volume 3).

The high cure shrinkage of polyesters is offset, to some extent, in products which have a high filler loading. Additionally, this shrinkage may have little if any consequence for some applications and in others it can be tolerated or accommodated. For example:

- void/gap and pore filling where adhesion is not critical (or where only compressive loads are imposed)
- applications where adhesion is important, but where gap sizes are relatively small (up to only 5–10 mm) and where surface roughness gives some mechanical key
- pre-casting
- thick linings and renders which are essentially self-supporting (particularly with fibre reinforcement).

Vinyl ester resins

As discussed above, unsaturated polyester resins contain certain reactive groups and repeating ester linkages in the main polymer chains. After cross-linking, the ester links remain together with some residual reactive groups. Both of these are potential sites for chemical attack such as hydrolysis and oxidation (although chemical and environmental resistance is still good).

The structure of the polyester chain can be varied at the resin manufacturing stage. One particular variation gives resins where both the ester and the reactive groups are confined to the ends of the polymer chains; the description "vinyl ester" arises from the particular chemical structure of these end-groups. Improved chemical resistance then results from the reduced number of "weak" groups in the cross-linked network.

As with the polyesters, styrene is usually employed as a resin solvent/cross-linking monomer and cure is promoted at ambient temperatures using similar initiators and accelerators; selection of the appropriate types and concentrations is again critical for optimum overall performance.

The longer chain segments between cross-linking points in vinyl esters usually results in greater toughness compared to the more densely cross-linked polyesters. This toughness, and the particularly high chemical resistance of vinyl esters, leads to their use in a number of more specialised applications where such properties are essential.

These improvements have to be judged against a higher cost. However, in critical applications involving protective linings/renders, or pre-cast products, for example pipes carrying very corrosive materials, good performance and a long service-life will usually be of primary concern.

Some vinyl esters, such as the urethane-acrylate type, have a lower cure shrinkage than the more standard vinyl esters (and polyesters). Hence, they can be used without difficulty in tough floor-toppings and mortars.

When vinyl esters are required, it will usually be necessary to seek out more specialist suppliers as proprietary products are relatively uncommon.

Reactive (meth)acrylate/acrylate resins

By using certain monomers, linear thermoplastic polymers with ester side-groups can be obtained, ie these groups are not a link in the main polymer chain. Two very common monomers of this type are methyl acrylate and methyl methacrylate although many others are available, allowing variation of the polymer properties. Methyl methacrylate monomer, produced on a large scale since 1927, provides the raw material for moulding compounds and for the manufacture of plastic sheet. Other thermoplastic (meth)acrylate polymers also have many direct applications.

In the early 1960s multi-functional (meth)acrylate monomers and pre-polymers became available. When these are polymerised with the more simple types, a cross-linked thermoset is obtained. In contrast to polyesters, the network now has a structure where the ester groups are mainly confined to a pendent position, ie they are linked to, but are not within, the main chains.

Thermosetting (meth)acrylates have been used increasingly in recent years, particularly for injection/impregnation, coating, patching and for forming floor toppings and waterproof membranes. These systems usually consist of a viscous syrup, containing the (meth)acrylate monomers, cross-linking agents and an accelerator, to which a filler/peroxide blend is then added. In general, (but with the exception of waterproof membranes) these resins are less widely used in the UK than in the USA and in Europe.

As the basic cure mechanism is very similar to that of polyesters, these two resin types have many resemblances, such as rapid cure at low temperatures. However, some significant differences can affect selection with the following properties supporting the use of (meth)acrylates:

- overall cure shrinkage is relatively high but this has little effect on performance because the major proportion occurs when the resin is still in the liquid and low modulus solid state. Higher stress relaxation also contributes as does the capacity for high filler loadings conferred by the low viscosity. Thus (meth)acrylates can be utilised for mortars, grouts, floor-toppings, coatings and lining compositions where good adhesion and substrate integrity is essential
- lower viscosity (for good penetration of porous solids and high filler loadings)
- greater toughness
- good alkali resistance
- improved outdoor durability and colour stability
- good intercoat/interlayer adhesion after full cure – fairly unique amongst reactive polymers which, in general, have unreceptive surfaces. With (meth)acrylates, the monomer softens and diffuses into the cured resin, thus giving a "weld" at the old/new interface. This has particular benefits in flooring (maintenance of in-service toppings; sealing and protecting filled systems using low viscosity (meth)acrylate coatings).

To benefit from these characteristics, the following disadvantages have to be accepted (in comparison to polyesters):

- higher cost
- a strong odour
- lower resistance to attack by some solvents
- increased fire risk (a flash point of circa 10°C, as against 32°C for polyesters).

2.4.4 Polyurethanes

The commercial development of polyurethanes, which dates from the late 1930s, has led to their wide use in many adhesives, flexible and rigid foams, surface coatings and elastomers. Although used in mortars, and more extensively in flooring applications, polyurethanes are most commonly encountered as coatings and elastomeric sealants for movement joints.

The chemistry of polyurethanes and their precursors is complex; a great variety of starting materials can be used to form the cross-linked structure and many side reactions can occur. However, most commonly, polyurethanes are formed from a diisocyanate and a polyhydric compound (a polyol).

These polyurethanes are supplied as two-pack products requiring careful proportioning and thorough mixing, as with epoxies.

Single-pack polyurethanes are also available and, here, the cross-linking cure reaction is induced by atmospheric moisture. Carbon dioxide is slowly evolved during this reaction. Consequently, the presence of excess moisture and/or too thick a layer must generally be avoided to prevent the formation of voids in the cured polymer.

Potentially, polyurethanes have even greater versatility than epoxies, especially as very flexible and tough products can be readily formulated. However, this potential has not been translated into a particularly wide range of products for construction uses.

Consequently, with only a few exceptions, this generic type is likely to be considered for only two applications, ie high performance coatings/membranes and floor toppings. Here, selection in preference to other resin types will generally depend on the specific service details.

Polyurethanes are frequently employed as high performance elastomeric joint sealants (single pack) and as foamed products (including systems for injection into cracks and voids). These applications are outside the scope of this report.

The relative lack of exploitation of polyurethanes may be partly due to the earlier establishment of satisfactory product ranges based on other resins, epoxies in particular. However, the following characteristics may also contribute:

- sensitivity to moisture during cure
- relatively high viscosities
- the highly toxic nature of some diisocyanates (although hazards are reduced markedly by using pre-polymers rather than the more volatile monomers).

2.4.5 Furane resins

Unlike the resins considered earlier, which are mostly synthesised from petrochemicals, furane resins are derived from agricultural residues such as corncobs and rice and oat hulls. These yield furfuraldehyde (furfural) which is converted to furfuryl alcohol.

The dark, free-flowing liquid resin used in furane compositions is then obtained by the controlled polymerisation of furfuryl alcohol to a pre-determined degree of conversion at a high temperature (circa 100°C). The resin obtained from this process (occasionally supplied in a ketone solvent) may then be cross-linked at ambient temperatures using an organic acid catalyst.

Although not extensively used in construction, furanes have particular benefit where extreme resistance to abrasion, heat and chemical attack is required, for example in rigid jointing and lining applications. Strength is usually retained above 100°C (above 200°C following post-cure) and resistance to hydrofluoric acid and hot concentrated alkali solutions can be obtained.

Furanes rarely appear in common product ranges. They are speciality resins whose use is limited to several applications where their outstanding chemical resistance and elevated temperature performance is essential (and where selection is usually obvious), such as:

- pre-casting
- bonding and grouting of acid resistant bricks and tiles
- rigid jointing of clay pipes carrying very corrosive chemicals and effluents
- linings and coatings for very aggressive conditions.

High cure temperatures (or post-cure) are necessary to maximise the performance of furanes. When applied to alkaline surfaces, eg concrete, a special primer is essential.

2.5 SOME CONSEQUENCES OF REACTIVE POLYMER CHEMISTRY ON APPLICATION AND SERVICE

When thermosetting resins are used a chemical reaction is being carried out. This places certain demands on the user. A number of essential requirements, all with a direct chemical origin, are summarised in Table 2.4 and discussed in the following sections (2.5.1–2.5.4). Further elaboration is given in Volumes 2 and 3 of this report (CIRIA PRs 78 and 79).

Table 2.4 *Requirements in use*

Correct proportioning of components	1. Do not formulate on site 2. Purchase pack sizes appropriate to needs 3. Discard leaking packs 4. Do not split packs unless specified 5. Re-disperse settled-out constituents 6. Drain containers thoroughly 7. Avoid spillage
Thorough mixing	1. Use recommended equipment 2. Ensure mixing is effective throughout the mass 3. Allow sufficient time
Temperature limits	1. Use suitable grades for the climate 2. Work only within recommended limits 3. Use controlled storage when close to temperature extremes 4. Consider the effects of exotherm and air/substrate temperatures
Adverse materials	1. Discuss unfamiliar substrates/environments with suppliers 2. Do not incorporate additional materials 3. Use clean mixing equipment
Health and safety*	1. Read supplier's information sheets 2. Note special precautions arising from local conditions

* See also: Volume 2, Section 3 and Volume 3, Section 3.2.3.

2.5.1 Proportioning of components

All chemical reactions require some proportioning of starting materials to achieve an optimum reaction rate and yield of the desired end-product. Thermosets have such a complex and extensive structure that normal chemical representation of a unique molecule is not possible; the material is defined by the type of units in the structure and the density of cross-linking. On this basis, it could be assumed that there is considerable leeway in the mix; this is not so. Some variation in the final structure may not significantly affect properties. However, there is generally a strict limit to this variation and, therefore, to the tolerance in proportioning the starting materials.

A hard product with a satisfactory appearance may well be obtained from a poorly proportioned epoxy or polyurethane. However, properties such as chemical resistance, modulus, creep and adhesion may be downgraded severely because the cross-linked structure deviates too far from that intended.

To obtain a structure within acceptable limits, the formulator defines the optimum blend ratio of starting materials (by weight). This is determined experimentally and, also, on a theoretical basis using the specifications of the raw materials.

Once this mix ratio is confirmed, products can be (and normally are) supplied in pre-weighed packs to facilitate correct proportioning. It follows that appropriate care is then demanded of the user, ie:

- complete transfer to mixing vessels must be achieved with no spillage
- supply containers must be drained completely
- packs should not be used if leakage has occurred
- with a few exceptions (see below) packs should not be split on site –proportioning correctly on a volume basis is unlikely due to liquid residues in measuring vessels.

Where proportioning from bulk, or pack splitting, can be carried out on site, this should be clearly stated by the supplier. Two typical examples are given below.

1. Some highly solvented epoxy systems used, for example, to impregnate concrete. Here the very low viscosity allows accurate batching by volume (usually in simple proportions).
2. Undiluted epoxies used for resin injection where purpose-built proportioning/ injection equipment is used.

Such cases are exceptional and complete packs of epoxy (and polyurethane) systems should normally be mixed.

Pack splitting is more commonly allowed for filled polyester and (meth)acrylate systems as slight variations in the mix ratio may be tolerated. However, care is still essential and the suppliers' mixing instructions must be followed.

Proportioning and blending may be carried out in factory operations where special formulations are required and where weighing or accurate volume metering equipment is available.

On-site proportioning frequently leads to problems and failures. If specific instructions for pack-splitting are not provided, then only complete units should be mixed.

2.5.2 Mixing of components

Where the cross-linked structure is formed from two initially separate components, ie epoxies, furanes and 2-pack polyurethanes, very thorough mixing is of paramount importance. When mixing is carried out poorly, solidification is still likely but further cure will be severely limited as active groups become isolated. For reaction to continue in the solid state, where long-range molecular movement is limited, a uniform distribution of reactants must be imposed at the mixing stage.

The peroxide catalysed type of resin (unsaturated polyester, vinyl ester, (meth)acrylate) is again somewhat more tolerant as all the network components are intimately mixed from the outset. However, thorough mixing of the resin and catalyst should still be implemented to enable cure to proceed uniformly throughout the mass.

2.5.3 The effects of temperature on cure reactions

Temperature has a marked effect on all chemical reactions, the rate of reaction often doubling with an approximate 10°C temperature increase. If the temperature of the reactants is lowered sufficiently, reaction will cease altogether due to the lack of internal energy. Reaction will usually continue if the temperature is subsequently increased although a lower degree of cure may result (see also Volume 3).

These effects must always be considered when using thermosetting systems. Manufacturers generally recommend an application temperature range which is consistent with cure proceeding satisfactorily at the lower extreme, while allowing adequate time for placement at the upper temperature. As such recommendations usually refer to the materials themselves, potential influences of air and substrate temperatures must be taken into account (see also Volume 2).

Particular care is required at the higher temperatures. Heat is often generated in the mixing process and considerable heat can be evolved during the reaction. These effects, especially marked with large masses and lightly filled resins, can lead to premature gelation and, consequently, insufficient working time (strictly, the quoted useable life should be related to a mass or volume). Pre-cooling can assist in such cases and initial overnight storage in air-conditioned enclosures can generally be recommended when working in hot climates.

At the lower extreme, it must be borne in mind that most resin systems have a low thermal conductivity. Consequently, if resins are stored outside in winter conditions their temperature may be very low, although the ambient temperature is acceptable. Overnight storage in heated surroundings will assist reactivity and give a lower, more useable viscosity.

In general the lower application temperature for epoxy and furane systems is approximately 5–10°C and for polyurethanes about 0°C. Provided an appropriate catalyst/accelerator system is used polyesters, vinyl esters and reactive (meth)acrylates can be cured satisfactorily at temperatures down to approximately -10°C to -20°C. Due to the wide variety of systems available, the specific limits should always be checked with the supplier.

To facilitate use over a wide range of ambient temperatures, specific products are often supplied in different grades; thus, a suitable useable life and cure time can be provided for tropical and much colder climates with little, if any, variation in service properties.

2.5.4 Adverse materials

Many polymerisation reactions require a carefully controlled environment which can only be provided in a laboratory, or chemical plant. Fortunately, most resin cure reactions are less demanding, allowing use under diverse conditions on a very wide range of substrates. However, some attention is still required: mixing vessels should be clean and further addition to the system, such as fillers, pigments or solvents, should not be attempted without detailed reference to the supplier. Although tests by the user may show that modification has no detrimental effect on the useable life, and that the resin still hardens, full cure may not in fact be attained and, thus, critical properties may be adversely affected.

The following common substrates may interfere with the cure reactions and/or adhesion:

- galvanised surfaces
- preservative treated timber
- bituminous products
- cured epoxies to which polyesters are applied (as residual amine hardeners can destroy the peroxide catalysts)
- alkaline contaminants and surfaces, eg concrete, to which furanes are applied (as the organic acid catalysts can be neutralised).

Special barrier primers may be available to combat such problems.

In cases where a specific surface is not referred to in a product data sheet, advice from the supplier should always be obtained.

Atmospheric oxygen can affect the cure of (meth)acrylates and some polyesters and vinyl esters – this is often manifested as a persistently tacky surface. A small level of a wax, which acts as a surface barrier, is incorporated in (meth)acrylate systems to avoid this problem.

Some cure reactions are sensitive to moisture although specific products are available for underwater applications. A clear distinction must always be drawn between cure and adhesion in wet conditions – good cure does not guarantee satisfactory bond strength.

The components of resin products, particularly solvents, can attack some substrates, eg some plastics; in other instances, staining may occur.

2.5.5 Health and safety

The hazards associated with resin materials are well known (see Section 3, Volume 2). However, the growth in concern with health, safety and environmental issues and the resultant legislation now dictates that the raw material supplier, the formulator and the user pay close attention to these matters. In all areas of industry, the use of some substances, hitherto quite common, has become increasingly restricted as information on specific hazards has become available.

Thus, established resin formulations have to be modified if it is found that certain components are unacceptable. Inevitably, the value of some long-term performance data and experience may then decrease as new products evolve. A reluctance to invest in further costly evaluation may ensue where additional legislation is possible. Additionally, for the highest levels of safety, some compromise in performance may have to be accepted; "the adequate" rather than "the best" may then become acceptable. Alternatively, uncommonly stringent precautions could have to be taken during use.

Health and safety risks, including those associated with fire and explosion, should always be assessed with reference to safety data sheets and product labels.

Many resin products contain components which may lead to both acute and chronic systemic effects via ingestion, inhalation or skin absorption, including respiratory problems, eye damage and dermatitis. However, with proper precautions and handling procedures, the potential dangers can be eliminated or reduced to an acceptable level (see Section 3, Volume 2).

Risks of fire and explosion can arise directly from product components (see Section 3.3, Volume 2 and Section 4.5.3, Volume 3) and, also, from the exothermic reaction of large batches (see Section 2.5.3 above and Section 4.6.2, Volume 3). In general, these risks can be reduced by:

- separation of storage and work areas
- good ventilation
- the elimination of ignition sources (flame-proof equipment can be required in some cases)
- conforming to supplier's recommended limits regarding batch size and/or section thickness.

3 A survey of the applications of reactive polymers in construction

3.1 INTRODUCTION

This section provides an introduction to the wide use of reactive polymers in construction. Many of these applications have become areas of specialist knowledge and activity. Where greater detail is required, reference should be made to the Bibliography.

Reactive polymer-based products are high performance materials. The relative importance of specific properties, such as strength, chemical resistance, viscosity or adhesion, varies from one application to another. However, most uses can be grouped under a few key words which convey distinct requirements. Such a classification is used to provide an overview in Figures 3.1–3.5.

Anchoring
- rock bolts
- fixings and holding down bolts
- ground anchors
- cavity wall ties

Segmental construction

Externally bonded reinforcement
- steel
- frp

Tiling and brick slips

Resin injection
- de-bonded renders and tiles
- externally bonded reinforcement

Miscellaneous
- stained glass
- road studs
- pile splicing
- rebar connectors
- pipe jointing
- rubber bandage waterproofing
- steel
- timber
- composites
- new to old concrete

Figure 3.1 *Adhesive applications*

Tiling and brick linings

Support/bearing
- machinery bases
- crane rails
- bridge bearing pads

Bedding
- man-hole frames
- kerb-stones
- runway lights

Sleeve/jacket repair (encasement)

Figure 3.2 *Grouting and gap/void filling*

Concrete*

- patching
- resin injection
- rebar coating
- protective coating and waterproofing
- rendering/lining
- polymer impregnated concrete

Modification of Portland cement concrete and mortar

Timber

- patching
- consolidation
- bonded dowels

Masonry/brickwork

- consolidation
- waterproofing
- lining/coating

Metals

- coating

* Pre-cast and reinforced elements – beams/columns/panels/pipe and tunnel segments/tanks. Damage due to reinforcement corrosion/cracking/impact and blast/abrasion/chemical attack.

Figure 3.3 *Repair and protection*

Flooring

- toppings
- coating/sealing and waterproofing
- resin injection of delamination toppings
- joints
- damp-proof membranes
- patch repair

Bridges/roads/airfields

- patching
- overlays
- pic
- non-skid surfacing
- waterproof membranes

Nosings

- roads/bridges
- floors
- stairs

Figure 3.4 *Wearing surfaces*

Structural

- machine bases/frames
- pipes, tunnel segments, tanks, ducts, sumps and chambers
- lintels
- level crossing units

Waste encapsulation

Semi/non-structural

- simulated stone
- man-hole
- frames/covers
- drainage troughs, gulleys
- cills, walls, panels
- wall/floor/roof tiles
- sanitary-ware
- street furniture

Figure 3.5 *Pre-casting*

This classification has not been adopted in the following review as certain techniques warrant a more unified treatment, such as resin injection, coating, impregnation and patch repair. The applications are therefore discussed in a sequence which approximately follows the increasing consistency (or filler loading) of the products utilised. This order is also used in the tabulated selection guidelines given in Volume 3 (CIRIA PR78).

3.2 SURFACE TREATMENT

3.2.1 Introduction

This section deals with resin systems which are applied in various forms to the surface of many substrate types for one, or a combination, of the following reasons:

- to give protection to the substrate (chemical and/or physical)
- to provide a barrier against liquids, gases or vapours
- to enhance, or maintain appearance
- to improve safety in use.

These reasons are further divided in Table 3.1 where typical examples are also given. A classification of these treatments is given in Table 3.2, together with an indication of the resin types which are commonly utilised. This classification provides the basis for the following sub-sections.

Powder coatings, increasingly applied on site, are outside the scope of this report and, therefore, are not discussed in any detail.

3.2.2 Surface sealing using solvented and non-solvented penetrants

Composition and areas of use

Highly solvented penetrants generally have a long pot-life and a very low viscosity due to their low resin content (typically 10–30 per cent) and are most commonly based on epoxies (two-pack) or polyurethanes (two-pack or single-pack moisture curing). They are usually supplied as clear systems, free of solid pigments or extenders and are true solutions, not dispersions; thus, they readily penetrate porous substrates. As the solvent evaporates, the resin is deposited and continues to cure, giving a sealing effect via a pore-blocking mechanism. Thus, in contrast to coatings, good surface adhesion is not essential.

Such products may be used to:

- improve the abrasion, impact and chemical resistance of concrete, to prevent dusting and to ease cleaning (particularly of light-duty floors where existing non-slip characteristics are relatively unaffected by the sealer)
- reduce the ingress of water, oils, for example into concrete
- improve the durability of stone, particularly in environments heavily polluted by exhaust emissions and acid rain (for example see Sasse, 1990)
- consolidate timber
- prime and bind friable surfaces prior to over-coating or patch-repairing (usually with a similar generic type of material).

They are particularly useful where concrete upgrading is required but there is a wish to avoid conventional surface coating. This may be due, for example, to relative costs and ease of application, non-slip requirements or to aesthetics. Although the change in surface appearance is minimised by using such sealers, concrete does usually become somewhat darker with a slight surface sheen or glaze. This difference is more noticeable when adjacent and untreated concrete becomes wet.

Table 3.1 *The need for surface treatment*

Need	Examples	Common substrates
To prevent direct degradation due to chemical attack	Reactions between concrete and mineral or organic acids, sulphates or other aggressive media.	Tanks, floors, pipelines, piles, tunnels, bridge piers, plinths, culverts, sumps, bunds, sewers.
	Reactions between metals and aggressive media.	Tanks, pipelines.
	Atmospheric corrosion.	Lattice and other structural steelwork. Timber.
	Breakdown due to moisture, ultra-violet light and micro-biological action. Degradation due to atmospheric pollutants.	Limestone.
To prevent direct degradation due to physical processes	Freeze-thaw induced damage of porous materials.	Concrete, masonry and clay-derived materials.
	Impact, abrasive wear and dusting.	Concrete floors and other trafficked surfaces, concrete subject to erosion by water flow.
To prevent indirect degradation due to chemical/physical processes	Loss of concrete integrity following carbonation or the ingress of chlorides and the subsequent corrosion of mild steel reinforcement or pre-stressing tendons.	Reinforced or pre-stressed concrete columns, beams, panels, bridge or car parking decks, tanks, pipes.
To limit the ingress or contact of liquids, vapours and gases	Waterproofing	Floors, basements, tunnels, culverts, bridge-decks.
	Barriers against moisture vapour, methane or radon.	Floors, basements.
	Handling potable water.	Water treatment plant and supply lines.
	Ease of cleaning and decontamination.	Floors and walls in hospitals, nuclear power stations, and areas connected with food/beverage handling, aircraft and automobile maintenance.
To enhance or maintain appearance	The prevention of mould growth and general dirt pick-up and staining.	Floors, walls.
	Anti-graffiti coatings.	Walls in subways and surfaces in other public areas.
	Uniformity of appearance.	Concrete following patch repairs.
To improve safety	Improved non-slip properties.	Floors and other foot-trafficked surfaces.
	Anti-static and electrically conductive surfaces.	Flooring in areas where hazardous materials or electronic equipment are present.

Table 3.2 *A classification of reactive polymer based surface treatment (by thickness)*

Classification	Resin types commonly used	Section
Surface sealers and impregnants:		
solvented	Epoxy, polyurethane	3.2.2
non-solvented	Epoxy, polyester, (meth)acrylate	3.2.2
Primers and surface coatings (0.1–1.0 mm)	Epoxy, polyurethane, (meth)acrylate	3.2.3
Thick membrane (1.0–5.0 mm)	Epoxy, polyurethane, (meth)acrylate	3.2.4
Linings (1.0–5.0 mm)	Epoxy, polyurethane, polyester, (meth)acrylate, vinyl ester, furane	3.2.5
Renders (≥ 5.0 mm)	Epoxy, (meth)acrylate	3.2.5

Notes

1. Floor toppings, overlays and non-skid road and deck surfacing are omitted from this classification.
2. A clear distinction between coatings, membranes and linings is not always possible. In this classification, coatings and linings are distinguished according to thickness; membranes refer to "sandwiched layers".

The use of solvented penetrants to achieve improved chemical resistance or a very high resistance to water ingress needs to be considered with some care. To ensure effectiveness, a multi-coat application is likely to be required – to the point where a distinct surface film is formed. Such performance may be achieved more easily and reliably with a product of higher resin content, ie a system still described as a sealer as some penetration is achieved or a conventional coating which predominantly gives a film on the surface. (Note that there is no rigid or formal criteria for the classification of surface sealing products – the division used here is mainly one of convenience. Thus, for some substrates, non-solvented systems may give a high degree of penetration; in other cases, even highly solvented systems may give a measurable film thickness.)

For very porous substrates, deep penetration of the solvented sealer will reduce the concentration of resin solids in the immediate surface zone – a high overall application rate may therefore be necessary to achieve the desired up-grading (with due allowance for drying between coats).

Where products with a very low viscosity can be readily formulated, eg polyesters, (meth)acrylates and some epoxies, solvent-free systems may be used as penetrants. Deep penetration, assisted by pressure application, can then be utilised beneficially as the potential problems of solvent retention do not exist. Here, the essential aim is to achieve a distinctly higher level of performance which may often include structural upgrading, eg the consolidation of old brickwork. This type of application is discussed further for concrete in Section 3.4.

Application

It is essential that substrates are dry, and in the case of cementitious materials well cured, in order to achieve an effective pore-blocking action. The application to polymer-modified mortar and concrete needs special consideration to ensure compatibility and effectiveness. When sealers are to be applied to new concrete floors, a lightly floated finish is preferable to heavy trowelling.

For upgrading defective timber, repeated application to refusal is usually carried out via small holes drilled at close centres. This treatment is regarded primarily as cosmetic and preservative, rather than as a structural repair. Although the strength of the timber can be increased considerably, this is difficult to quantify reliably (Bowden, 1993).

These sealers are easy to apply and generally require a minimum of surface preparation (if any). The high solvent level demands that close attention be given to health hazards and fire risk, particularly where natural ventilation is poor. Such problems can be eliminated by the use of two-pack water-based epoxy systems. However, these products are dispersions, rather than solutions, and penetration is generally more limited.

Alternatives

A number of alternatives to two-pack solvented resin sealers are available, for example:

- single pack solvented epoxies, polyurethanes and acrylics, ie solutions of thermoplastic polymers
- water-based dispersions of various thermoplastic polymers, eg acrylics, vinyls, PVA and SBR
- silanes, siloxanes and silicones
- silicates and silicofluorides.

The first two categories given above also function by a pore-blocking/film formation mechanism. Silanes and siloxanes (of which there are many variations) generally react with the substrate, eg concrete or masonry, imparting water-repellency to the pore surfaces while maintaining an essentially open and "breathable" pore structure. Silanes are specified in the UK for the protection of reinforced concrete highway structures against chloride ingress (Department of Transport, 1990). Silicates and silicofluorides are only suitable for concrete substrates as they provide pore-blocking via reaction with the free lime.

Performance

Concise guidance on the absolute or relative performance of solvented sealing systems is not given here as the products are so varied and the essential needs vary with the type of application, for example:

- timber consolidation – preservation and strength increase
- sealing exterior surfaces – good appearance, colour stability (non-yellowing) and resistance to water ingress and weathering
- priming friable, porous surfaces – a good adhesive connection between the substrate and top-coat or patch repair.

Generally, therefore, selection is based predominantly on previous experience and discussion with suppliers (possibly assisted by site trials and specific testing) rather than assessment against published standards.

Further information is available on the use of various sealers for the improvement of concrete floors, including some comparison of abrasion resistance where reactive resin-based systems were shown to give a superior performance (Shaw, 1994; The Concrete Society, 1994; Kettle and Sadegzadeh, 1986, 1987).

3.2.3 Surface coating

The range of products and selection

This application is one where resin-based products and their alternatives are particularly numerous and wide-ranging. For example, the *Code of practice for protective coating of iron and steel structures against corrosion* (BS 5493: 1977) considers approximately 80 complete protective systems.

It has evolved due to the following factors:

- surface coating is a long-established and major industry which is able to utilise many types of material
- general performance criteria, costs and substrate types can vary widely
- seemingly minor aspects of local application and service conditions can markedly affect performance, thus generating many very specific needs.

Due to the resultant complexity, this review does not include detailed selection guide-lines. However, as shown in the Bibliography, considerable guidance on surface coating has been published, including separate standards dealing with selection, testing and application. Effective selection, based on this literature and discussions with suppliers, demands that individual needs are defined comprehensively and precisely.

For certain applications of resin-based products, a service life equivalent to that of the associated structure or element (or any alternative material) can be anticipated, eg pre-cast products, load-bearing grouts, adhesives used in segmental construction. With most coatings, maintenance is virtually inevitable and the "life to first maintenance" (particularly of the more exposed or vulnerable areas) is an important basis for selecting a level of performance. Consequently, detailed lifetime cost analysis often becomes particularly critical in a selection process where, especially for a long life in demanding conditions, judgement and experience are usually required to supplement the more formal published guidelines.

The more common types of coating are summarised in Figure 3.6 thus showing resin-based products in a broader context.

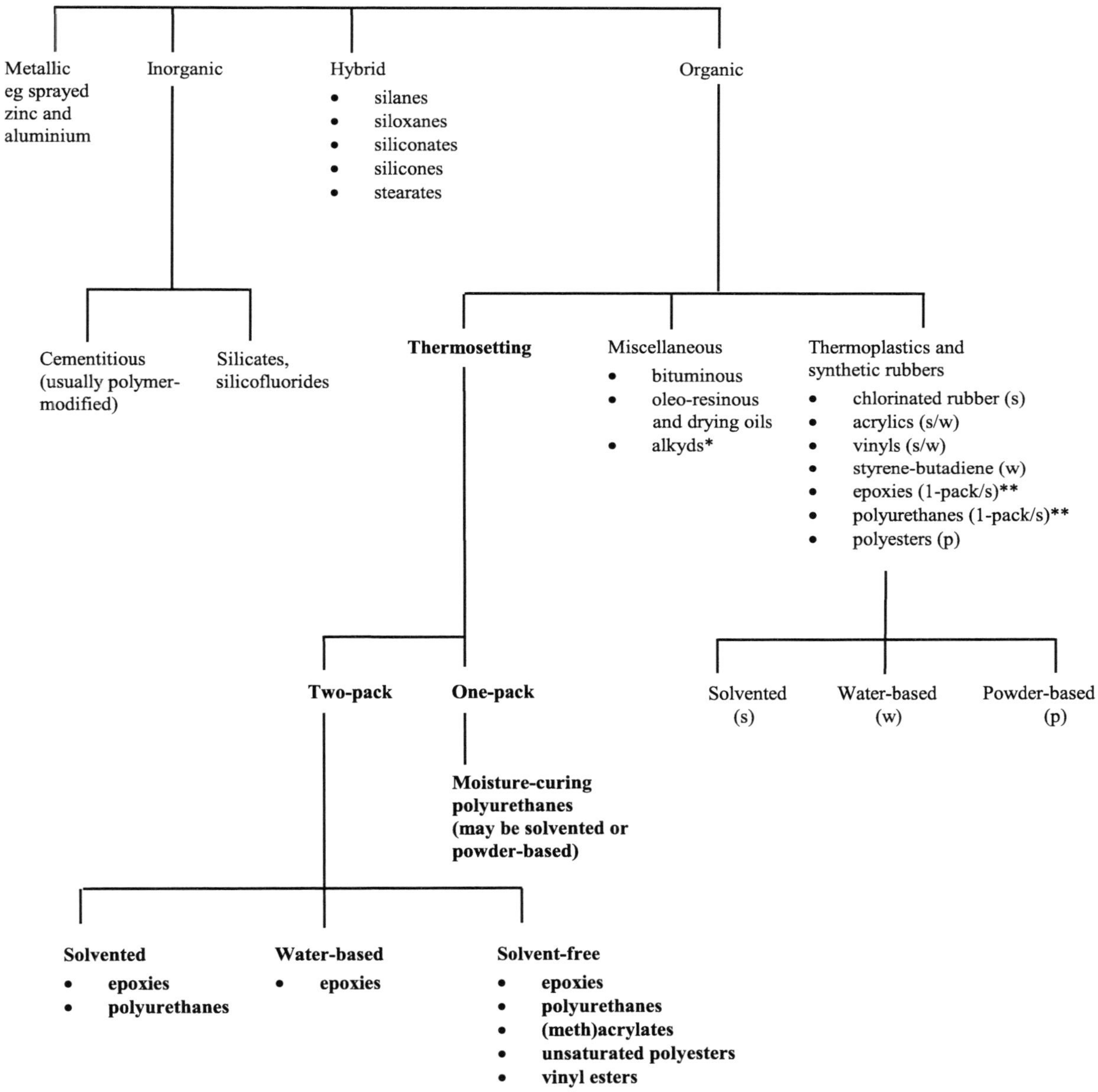

* Including urethane modified materials which may be referred to as "polyurethanes".

** Non-reactive – solid or high viscosity linear polymer in solution.

Figure 3.6 *A classification of surface treatment and coating systems*

A number of variations of resin-based coatings which warrant particular mention are summarised in Table 3.3.

Table 3.3 *Variations of resin-based coatings*

Variation	Comment
Zinc-rich primers	Two-pack epoxies where the very high content of zinc dust provides inhibition of steel corrosion by a cathodic protection/ sacrificial mechanism.
Blast and prefabrication or holding primers for steel	Based on special zinc-rich epoxies or two-pack epoxies containing iron oxide possibly combined with corrosion inhibiting pigments.
Coatings incorporating lamellar materials	For improved durability and barrier performance. Examples include some micaceous iron oxide pigments and glass flakes.
Tar and pitch extended epoxies and polyurethanes*	For good performance in marine and other wet service conditions at a reduced cost relative to the non-extended coatings.
Hard, impervious polyurethanes	Increasingly used as anti-graffiti coatings where good resistance to impact, abrasion and cleaning solvents is required in addition to good colour stability and ageing characteristics.
Non-slip/non-skid systems for lightly trafficked surfaces	Particularly useful for surfaces subjected to wet conditions, eg, floors, ramps, parking decks. A hard-wearing angular grit or fine aggregate (silica sand, calcined bauxite, aluminium oxide) is usually scattered onto the first wet coat, surplus being removed prior to application of a second binding coat. The resultant texture may represent a compromise between improved non-slip characteristics and the greater difficulty of cleaning.
Coatings which are dressed, eg, with sand, to improve adhesion	Used where excessive delay between successive coats (or between application of a primer and a resin mortar) is unavoidable, eg rebars coated during the patch repair of concrete.
Coatings for contact with potable water	Mandatory requirements must be met.
"Surface tolerant" coatings (or mastics)	Designed for application to poorly prepared surfaces and old coatings. These high-build systems (up to 2–3 mm/coat), often based on epoxies or, occasionally, polyurethanes, can vary widely in performance.
Flexible "mastics" or "sealants"	Often based on epoxies and used to seal crevices and to encapsulate rivets, nuts and bolts, prior to coating.
Coatings with low electrical resistance	Can be based on epoxies. For use in areas where anti-static/ conductive surfaces are required.
Air-field and road marking paints	Require good resistance to fuels, oils and hydraulic fluids.

* Coal-tar and pitch extended coatings have a very long history of use and are still commonly available. However, for health and safety reasons, the coal derivatives are being replaced by alternative extenders, eg synthetic hydrocarbon resins and petroleum derivatives including bitumen.

The majority of the coatings referred to earlier can be applied on site. Of the many other coating types which are most commonly applied under factory conditions, mainly to steel, powder coatings may be especially noted. These are formed from pigmented polymers which are applied as dry powders and then fused by heat. Both thermoplastic (polyethylene, vinyl copolymers, nylons) and thermosetting (epoxy, polyester, acrylate, polyurethane) types are available.

One particular type of powder coating, fusion bonded epoxy (FBE) resin, is commonly used for the protection of steel pipelines and other hollow sections. However, due to the extensive and world-wide problems resulting from corrosion of mild steel reinforcement (see Section 3.10), it is increasingly considered and used for the future protection of rebars in new concrete structures. This application originated in the USA approximately 20–25 years ago as a means of improving the durability of bridge decks (where reinforcement corrosion is particularly prevalent due to the extensive use of de-icing salts and the absence of waterproofing membranes).

Standard specifications for FBE reinforcement are available in the USA (ASTM A775/ A775M-88) and the UK (BS 7295, 1990). However, premature corrosion has been observed on a number of bridges in Florida (Manning, 1996) and concerns have been raised by research workers in both the USA and Europe. This suggests that further investigation and service experience is required before the true long-term benefit and cost-effectiveness of FBE reinforcement can be defined unequivocally.

Performance

Organic coatings can be sub-divided in two ways.

1. According to whether or not a solvent (or water) is present, ie:
   - solvent-free coatings
   - solvented and water-based systems where a wide range of polymer contents is possible.

2. According to the mechanism by which the dry film is formed, ie:
   - by physical drying
   - by chemical reaction
   - by a combination of these processes.

Thermosetting resins may be distinguished from most other polymers used in organic coatings as they can give chemically-curing and solvent-free systems. These characteristics are responsible for the following benefits:

- with the incorporation of a thixotrope, a high-build can be achieved even on vertical and overhead surfaces (at least 0.3 mm/coat). Thus substantial protective barriers can be formed with no possibility of the solvent entrapment which can lead to rapid failure
- the cross-linked structure of the cured film confers excellent durability in very aggressive service conditions (where resistance to corrosion, chemical attack or abrasive wear is required). Polyurethanes and vinyl esters can provide particularly tough and impact resistance coatings
- a satisfactory cure rate and good film formation can be achieved under conditions which do not favour solvented or water-based coatings. For example: cold, humid or enclosed conditions where drying could be very slow; hot, windy conditions where rapid solvent loss can lead to application difficulties and solvent entrapment in the lower part of the film. Additionally, some epoxy resin coatings can be applied underwater
- very reactive, high viscosity systems, usually epoxies, can be spray-applied with equipment incorporating heaters, proportioning pumps and twin supply lines. This enables high performance coatings to be cured relatively rapidly in poor conditions.
- The health, safety and environmental problems caused by solvents are eliminated (although hazards due to the resin system may still be present – see Volume 2).

Solvented coatings of epoxies and polyurethanes also give excellent barrier, chemical resistance and mechanical properties at a lower film thickness. Good performance can thus be achieved at reduced cost when a solvent-free high build coating is considered to be unnecessary ie where the very highest performance is not required.

Certain epoxy resin systems are also used in two-pack water-based coatings. The potential of such coatings is generally considered to be under-utilised in the UK (although their use is not favoured by humid, cold conditions). Many types of pigment and extender can be incorporated and reduced gloss finishes can be readily produced (in contrast to the solvented/solvent-free systems). When appropriately formulated, these coatings can be applied to steel. They can also be applied reliably to "green" concrete, assisting cure by inhibiting moisture loss while subsequently providing a protective film (which can be easily overcoated for added protection or modified appearance).

Although coatings based on thermosetting resins are generally classed as high-performance systems, there are limitations which have to be considered. For example:

- they vary widely in their ability to bond well to damp or wet surfaces
- the compatibility of different coating types varies. For new work, a complete system should be obtained from one supplier; for maintenance painting, a number of site tests may be necessary before suitable preparation and products can be specified
- although providing excellent barriers to atmospheric environments, some epoxy and polyurethane coatings are particularly prone to discoloration and surface chalking
- most thermosetting coatings are barriers to the rapid transmission of water vapour. Although essential for some applications, this property may be detrimental in others
- a low level of water uptake or high resistance to chemical attack by the coating itself does not necessarily guarantee good in-service performance. Only small concentrations of water or chemicals, reaching an interface by diffusion, may be sufficient to cause degradation of the substrate and/or de-bonding
- the upper service temperature for coatings varies widely but will particularly depend upon whether the environment is dry or wet.

For these reasons, it is usually essential that fully detailed application and service conditions are discussed with suppliers at an early stage. Data sheets for coating products in particular should be used as an introduction which assists further enquiries.

Application

High performance cannot be equated automatically to tolerance of poor application conditions. To achieve satisfactory service, certain requirements may have to be followed stringently. For example:

- excellent long-term adhesion will generally only be attained with high standards of surface preparation; specification of a high-performance coating is often pointless if this cannot be achieved. On steelwork, this includes the removal of water-soluble salts (which can cause osmotic blistering) in addition to the more obvious contaminants and blasting to the recommended surface profile. Fairing or levelling coats may be required on rough concrete surfaces, particularly where blow-holes are present; special sealing primers may be needed for particularly porous, friable surfaces (see Section 3.2.2)
- dryness may be required beyond the immediate surface zone. As a considerable period can be required for freshly cast thick concrete sections to dry, overcoating requirements (eg on floors) should be fully discussed at the design stage
- some common surfaces require special primers, eg galvanised steel. Others, like polymer-modified concrete, should be discussed with suppliers prior to coating
- particular attention should always be given to the application condition limits (temperature, humidity) and to the corresponding cure times, especially when early exposure to water, chemicals or mechanical action is likely
- it is essential that a recommendation for the maximum interval between coats is obtained and observed (and adjusted, if necessary, for local conditions). For many but not all systems, excessive cure can lead to poor inter-coat adhesion (the maintenance of resin coatings requires special consideration for the same reason).

New developments

Such are the varied and demanding conditions where the use of protective coatings is essential that improvements in performance are constantly being sought, for example:

- faster drying and curing times, particularly at low temperatures
- increased tolerance of poorly prepared and damp/wet surfaces
- improved ease of application at high film thicknesses
- good long-term flexibility with low internal stresses
- improved anti-corrosion performance
- easier long-term over-coating.

Due to their formulation complexity and their application to large areas, coatings are probably more affected by health, safety and environmental legislation than most other resin-based products considered here. Some common components have now been eliminated or reduced in use (lead and chrome-based pigments, coal tar, pitch, solvents (specifically and in general) aromatic amines); conventional two-pack epoxies and polyurethanes have also become restricted in some countries, such as Scandinavia. Consequently, newer materials with reduced hazards are becoming available, eg epoxy and amine-functional acrylics, urethane-acrylics and hydroxy-acrylic/anhydride systems.

3.2.4 Water/damp-proofing membranes

Composition and areas of use

The term "membrane" refers here to a thick barrier which is usually, but not essentially, sandwiched and therefore protected by other materials. Epoxies, polyurethanes (often tar/pitch modified and moisture-curing) and (meth)acrylates are widely used, in competition with bitumen-based materials (hot-applied or emulsions) and pre-formed sheet products.

These resin systems provide excellent barriers against water migration and moisture vapour transmission (although slow diffusion will take place). They are often formulated to give tough products with high elasticity, even at low temperature, and good chemical resistance. They are easily and rapidly applied by brush, spraying, roller or squeegee, readily conform to surface irregularities and are seamless (de-bonding tapes are initially applied over any movement joints). Fairly rapid cure at low temperatures is usually provided to assist the continuity of work and good impact and abrasion resistance are required to prevent damage during back-filling or over-laying. Good adhesion eliminates "tracking" of moisture; compared to non-bonded sheet waterproofing materials, leakage can thus be traced more easily.

Typical areas of application include foundations, basements, floors, car decks, roof-terraces, pools, walls, and tanks. Solvent-free (or water-based) epoxies in particular may be used both as a damp-proof membrane and a resin bonding-aid between sub-floors and cementitious floor toppings (BS 8204: Part 3, 1993); sand is often applied to the final coat to provide a key.

Bridge-decks

The use of waterproofing membranes on reinforced concrete bridge decks is important. It has been estimated, for example, that repair costs for many bridges in the USA are equivalent to the first cost of the structure – initial waterproofing would have improved durability significantly at low (approximately 5 per cent) added cost (Macdonald, 1974).

Large-scale trials using skid resistant epoxy surfacing materials were first carried out in the USA in the mid-1950s (see Section 3.3). It was quickly realised that the prevention of chloride ingress was an important benefit for reinforced concrete bridges. However, little work of this type was performed outside the USA until the mid-1960s.

From the 1940s onwards, most concrete bridges in the UK had a protective and water-proofing wearing course (mastic asphalt). However, with the increasing use of de-icing salts in the 1960s, reinforcement corrosion became more likely as the bituminous surface gave inadequate long-term waterproofing.

Shortly after the commencement of a large-scale programme of bridge building, more stringent membrane waterproofing of all concrete decks on motorway and trunk road bridges became mandatory with the issue, by the Ministry of Transport, of Technical Memorandum BE1 in 1965. This was superseded in 1970 by TM BE27 and, following numerous amendments, a completely new standard for all bridges, BD47/94, was issued by the Department of Transport in 1994.

While BE1 and BE27 predominantly addressed the use of mastic asphalt and sheet waterproofing membranes, BD47/94 also deals with liquid applied systems which may include resin-based products. Other European countries, eg France and Germany, also have national regulations relating to bridge deck waterproofing.

On a large proportion of highway bridges and elevated highways, both in the UK and abroad, fast-curing epoxy, polyurethane and (meth)acrylate systems, commonly spray-applied, have become widely used as a waterproofing membrane between the deck and a wearing course (Potter, 1975; Price, 1990). Specific examples are provided by the Forth Road Bridge and the Bosporus I and II Bridges.

The new UK Standard BD 47/94 contains details of the certification procedure which came into force on 1 April 1996; this includes a range of laboratory performance tests and a field-trial. The handling and performance requirements include:

- resistance to pin/blow-holing
- setting times
- tensile and shear adhesion
- resistance to chloride ion penetration, freeze-thaw, ageing, chisel impact, and aggregate indentation
- thermal shock, ageing and crack cycling
- adhesion between membrane and deck surfacing.

In the UK since the 1970s, at least several hundred bridges have been waterproofed with thermosetting membranes which have the following attractions:

- very rapid airless spray application with consequent cost savings
- good cure and adhesion under cold damp conditions
- good adhesion to concrete, steel and bituminous deposits and good inter-coat adhesion
- tough, flexible and seamless membranes which can be formed on vertical detail, in addition to the horizontal surface, at 1–2 mm thickness/coat
- good chemical resistance and barrier properties.

Health and safety requirements are likely to eliminate the use of tar/pitch extended systems and particular precautions are needed during the application of polyurethane systems which contain free isocyanates.

3.2.5 Linings and renders

Areas of use

A much thicker layer than that provided by conventional high-build coatings is often desirable where surfaces (concrete and masonry in particular) are exposed to impact and abrasion in addition to water or aggressive chemicals. Where liquids are flowing at relatively high velocities, protection against cavitation damage can also be required. Such conditions may occur, for example, in tanks, effluent sumps, pits or channels, sewers and other pipes and the hydraulic spillways associated with water-treatment works, dams and hydroelectric power stations.

In these and similar applications, heavily filled mortars based on thermosetting resins may be considered as alternatives to fibre-reinforced linings (also based on reactive polymers but outside the scope of this report), high-performance cementitious products, pre-formed steel, plastic or rubber linings, engineering bricks or tiles.

The products used in such cases have some similarity to heavy-duty trowel-applied floor toppings although special attention has to be given to the formulation for vertical/over-head application and, where areas are very large, the effect of binder type and filler loading on costs.

In a number of applications, for example continuous linings in tanks and pipes, the resin system can act in a virtually self-supporting manner due to its thickness and high strength. However, it is normal to prepare substrates thoroughly to achieve good adhesion and thus to give assurance of maintained integrity, particularly where a back-pressure exists, eg from groundwater.

A number of applications closely associated with those discussed below utilise pre-cast resin-concrete and are discussed in Section 3.12.

Specific applications – protection against mechanical/chemical action

Some typical examples of this type of application, which is particularly common in Japan, Russia and the USA, are given below.

1. A number of uses on **hydraulic spillways** in Russia dating from about 1960 onwards have been summarised (Mikhailov et al, 1992). In one particular case, 100–150 mm thick resin concrete layers were applied to a total area of 900 m^2 on the spillways of five dams; over 25 years operational experience has demonstrated the economic benefit conferred by high wear resistance and durability.
2. The technology for **spray-applying epoxy mortars** has been developed in Japan to repair concrete structures, including tunnels used to carry sewage or irrigation water. This process (the "shot rem system") uses compact lightweight equipment which can be operated in confined spaces and which provides continuous mixing of the resin mortar. Losses, including rebound, are low (2–3 per cent) and, typically, a lining thickness of 5 mm can be applied at 40–80 m^2/hour (Sakamoto, 1984).
3. Vinyl ester mortars have been used for high performance **linings in hydroelectric power stations and areas handling fertilisers** in China (Lin et al, 1986).

4. Thick polyurethane linings for **pipes conveying abrasive materials**, either hydraulically or pneumatically, are reported to have a significant cost benefit compared to unlined steel pipes or rubber linings (Galka and Yates, 1984). Epoxy mortars are also used for the ***non-structural re-lining of water-pipelines*** giving good corrosion protection, rapid re-commissioning, a minimal loss of pipe diameter, a smooth surface and good durability (ibid). Application to wet surfaces is also possible with some systems.
5. The use of **furane linings for various types of concrete pipes** in Czechoslovakia has been described (Bares, 1978).
6. Various uses of **resin-based linings** in American installations concerned with the **transport of water** have been reviewed (Scanlon, 1981) as have applications in **geothermal plants** where resistance to hot brine and steam is essential (Zeldin et al, 1981). In the latter area, appreciable economic benefits over a 20-year operating life are claimed (Kaeding, 1985).
7. Good performance of **anti-cavitational furane linings** has been demonstrated at the Volga hydroelectric power station since 1960; epoxy-based materials used in similar installations in Russia have performed well for at least 15 years; brief reference is also made to the extensive use of epoxies in the USA and to studies carried out in France and other countries (Mikhailov et al, 1992).

For more information on similar applications, the published papers from the ICPIC Conferences should be consulted (see Bibliography).

Specific applications – thermal insulation, cathodic protection, surface levelling

Thick resin-based linings/renders have also been used where properties other than protection against mechanical action and chemical attack are required. Some typical examples are given below.

1. The incorporation of lightweight aggregates can give an **impermeable and thermally insulating mortar/concrete** for use on walls, roofs and other surfaces – particularly on retaining bunds for liquefied natural gas tanks where hazards are reduced by a lower evaporation rate following spillage (Mikhailov et al, 1992; Fontana et al, 1987 and 1989). Such materials have been applied by hand or by a special spraying process (see above) and are based on epoxies, polyesters or vinyl esters (proprietary lightweight epoxy mortars are commonly available in the UK and are often used for high-build patch repair). Two installations in New York utilised low modulus epoxy binders which withstand exposure to the weather, including thermal cycling. Strengths were fairly low (UCS of up to 25 MN/m^2; flexural of about 4 MN/m^2) but adequate for the loads imposed by vehicular and pedestrian traffic during tank service/maintenance. The LNG boil-off rates were reduced by up to 85 per cent (Mendis, 1987).
2. **Electrically conductive polymer concrete** based on polyesters and vinyl esters has been developed in the USA as a spray-applied thick coating for the vertical and overhead surfaces of reinforced concrete bridge sub-structures (Fontana, 1987). This coating forms part of an impressed current cathodic protection system for the reinforcing steel (an increasingly common technique for bridge decks and other reinforced concrete structures).
3. More routinely, epoxy mortars or pastes are used as **renders or "scrape-coats" to level concrete surfaces** prior to epoxy coating. The speed of cure and the subsequent absence of surface moisture are beneficial – often essential. However, cementitious materials may provide an acceptable and less costly alternative.

Structural applications

In certain instances, such as the three examples listed below, linings and renders may be used structurally.

1. **Resin concrete** with a thickness of 500 mm has been used in Russia **to support mine-shafts** driven in frozen rocks (Mikhailov et al, 1992).

2. **Pipe-lines and sewers** in poor condition and ranging in diameter from 100 mm to >1000 mm can be completely **renovated** using the non-man entry system, "Insituform", developed in the UK (Galka and Yates, 1984). This process employs a felt tube or bag which is sealed on one side with a polyurethane and then impregnated with an uncured but activated polyester resin. Water pressure is used to simultaneously turn the tube inside out as it feeds through itself and along the pipe, thus also forcing the resin against the pipe-wall; hot water is then used to cure the resin.

3. A 1200 mm internal diameter ***automated shield tunnelling system for telecommunication cable lines*** has been developed successfully in Japan (Matsuura et al, 1981; Kobayashi et al, 1981; Nojiri et al, 1981; Takatsuka, et al, 1984). This process utilises a low cure shrinkage polyester mortar which is capable of setting in wet conditions to give high compressive and flexural strength (30 MN/m^2 and 10 MN/m^2 respectively) within 30 minutes. The 100 mm thick lining is produced by an automatic slip-forming method with tunnel lengths up to 200 m being achieved in the trials reported in 1984 (Figure 3.7).

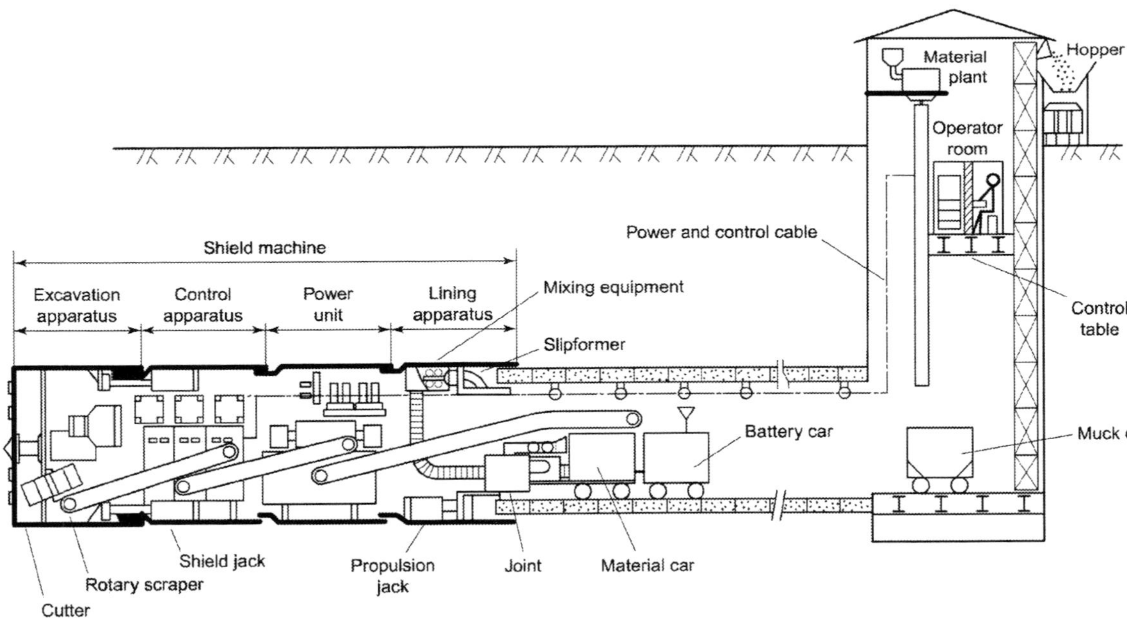

Figure 3.7 *Outline of automated shield tunnelling system for telecommunication cable lines (Takatsuka et al, 1984)*

3.3 SKID RESISTANT WEARING SURFACES AND PROTECTIVE OVERLAYS

Slip (and skid) resistant surfaces for foot and relatively light vehicular traffic are discussed in Sections 3.2 and 3.11. Resin-based systems are also used on the following areas where resistance to skidding is essential:

- car-parking decks and access ramps
- roadways where accidents due to skidding are likely, eg the approaches to junctions and roundabouts
- reinforced concrete bridge-decks.

In these cases, good resistance to wear by heavy trafficking is required and good adhesion must be maintained under varying weather conditions. In addition, for bridges and parking decks, a protective waterproofing function is usually necessary.

Proprietary systems based on polyurethanes, epoxies or (meth)acrylates are available for car-parking decks and similarly trafficked areas. These are generally thick coatings which utilise an appropriate aggregate (pre-mixed or broadcast) to give the skid resistant texture. They have good resistance to petrol and oils and, in conjunction with suitable drainage, the deck is waterproofed (thus assisting the prevention of reinforcement corrosion where de-icing salts are deposited by the traffic).

3.3.1 Skid resistant road surfacing

Historical background

The first large-scale road surfacing trials using epoxy resins were performed in the USA in the mid-1950s (Potter, 1975). The main aim of this work was to utilise the excellent adhesive properties of the resin to bond sharp, hard aggregate to surfaces that had worn smooth and thus to improve the skid resistance.

In the USA, the barrier properties of these compositions became increasingly important leading to the development and extensive use of skid resistant overlays primarily for the protection of reinforced concrete bridge decks (as discussed in Section 3.3.2).

In the UK and Europe, developments in the protection of bridge decks moved in a different direction ie, towards the use of waterproof membranes placed between the structural concrete deck and the bituminous wearing surface (see Section 3.2.4). However, the originally more dominant aim in the USA, the improvement of skid-resistance, was pursued in the UK for the purpose of reducing accidents on urban roads, particularly in wet conditions.

This need was identified on the basis of accident statistics compiled in London. It was shown that over 70 per cent of serious accidents occurred within 20 m of certain junctions and crossings and that a large proportion of these involved skidding vehicles, although the road surface appeared to be sound.

The development of a portable tester allowed quantitative measurements of skid resistance (Giles et al, 1964) to be correlated with accident statistics, confirming that surfaces were often inadequate – mainly due to rapid polishing of the road-stones.

While it was known that calcined bauxite in particular was extremely resistant to polishing under heavy traffic, this aggregate was too costly for use in asphaltic surfacing compositions. Consequently, adhesively bonding a single layer of bauxite to the asphaltic surface was considered a more economic alternative.

Small-scale trials in central London, in 1966, initiated widespread use in the UK. Such surfacing is now standard practice for certain sites and is included in the *Specification for highway works* issued by The Highways Agency (1991/1993). Similar development occurred in many other European countries.

Detailed accounts of both the early work and later developments are given in the reports issued by TRRL (now TRL) which are noted in the Bibliography (TRRL, 1974; Wright, 1978). Other useful descriptions have also been published (Potter, 1975; Durst, 1974; Hatherly et al, 1969).

Composition

Many epoxies and other thermosetting resin adhesives (and more recently hot-melt thermoplastics) together with various aggregates have been assessed (and, in a number of cases, used in the UK). Since the earliest trials, very many applications have employed particular 2-component bitumen-modified epoxies (with calcined bauxite). With one system the resin is applied in a hot state at about 1.0–1.5 kg/m^2 prior to broadcasting the aggregate. The final amount bonded after removal of the excess is about 5 kg/m^2.

This system is considerably less costly (by at least 50 per cent) than the skid resistant overlays discussed in Section 3.3.2 as the resin consumption is far lower, merely being sufficient to anchor the aggregate rather than to form a waterproof membrane. However, the general need to reduce costs while maintaining or improving performance, in addition to commercial competition, has led to the continued assessment of alternative binder and aggregate systems.

Considerable attention has been given to alternative aggregates as calcined bauxite is expensive. However, as the bauxite aggregate is extremely resistant to polishing and wear by abrasion and weathering, it maintains skid resistance at a high level for a considerable length of time, and, therefore, continues to be used.

Areas of use

Two factors contribute to the proven cost-effectiveness of epoxy resin surfacing systems:

- a durability of at least ten years under severe traffic conditions (where other types of finish have inadequate cohesive strength or long-term skid resistance)
- a considerable reduction in the number of skidding accidents (by at least 70 per cent in wet conditions).

This surfacing becomes less economic where the risks of skidding are relatively low. Alternative and cheaper means of achieving satisfactory skid resistance in such areas are available. Consequently, the resin-based surfaces have been applied mainly to "high-priority" critical sites such as approaches to roundabouts, pedestrian crossings and signalled junctions/intersections on major routes (in the UK. such sites are defined by quantitative standards for skid-resistance). However, flyovers and bridges, where skidding accidents cause major traffic blockages, have been similarly surfaced.

Application

Surfacing of 500–2000 m^2 is typical for roadways but on the Severn Bridge some 24 000 m^2 was treated. To cover these areas efficiently and quickly with a hot applied fast-setting resin, special truck-mounted metering, mixing and application equipment allows dressing of at least 5000 m^2 per day. The resin is usually applied automatically through a spray-bar. A large excess of aggregate is then scattered with a modified grit-spreader, the excess being recovered with a suction sweeper once the resin has cured.

The epoxy/bauxite system has been mostly used on relatively sound hot-rolled asphalt surfaces where only sweeping and the removal of very oily patches are necessary. The surface must be free of excess water although light condensation can be tolerated. Other surfaces, such as mastic asphalt and concrete, have been treated (Durst, 1974) but additional preparation may be required (and long-term performance may vary). For example, in resurfacing the Severn Bridge in 1991, it was found necessary to roughen the surface of the mastic asphalt by flailing, prior to application of the epoxy/bauxite system.

Properties

Fast curing is essential to minimise the period of road closure – generally the work is carried out at night and the road is re-opened for the morning traffic. Typically, sufficient cure was obtained with the early epoxy systems in less than two hours at 20°C and less than five hours at 5°C; more recent products give an even faster rate. Hot application, assisted by the automatic equipment, both improves the rate of cure and optimises the viscosity of the resin.

Other essential properties are:

- resistance to oil, fuels, water and weathering
- good adhesion under the shear stresses imposed by traffic
- a flexibility which is compatible with that of the road structure in order to avoid brittle or fatigue failure
- good cohesive strength and ability to retain the aggregate.

3.3.2 Skid resistant and protective bridge-deck overlays

Background and types of overlay

Concrete cracking, spalling and delamination induced by reinforcement corrosion is a major problem throughout the world, especially where chlorides come into contact with the steel. De-icing salt is, therefore potentially aggressive, particularly where poor concrete cover and/or quality facilitates penetration.

Reinforced concrete bridge decks (and the sub-structures) are prone to deterioration where de-icing salt is used extensively to keep roadways open during the winter. When required, repair is very difficult if the bridge is heavily trafficked.

In the UK and some other European countries, these problems have been largely (if not completely) avoided by the use of bituminous top surfaces or, at later dates, by waterproof membranes formed between the structural deck and the wearing course (see Section 3.2.4). However, in the USA, although winters are severe and a high proportion of bridges are constructed from concrete, such protection has been less commonly used and major repair and protection programmes have had to be implemented since the early 1960s.

Many publications refer to the use of "polymer-concrete" to form new wearing surfaces on bridge-decks, particularly in the USA (eg ICPIC and ACI Conference Proceedings). This description can be misleading, as the following two systems are not always clearly distinguished.

1. Formation in-situ by embedding or sandwiching an aggregate between successive applications of a resin-based coating or slurry (see below: Overlays built-up in-situ).
2. Direct application of a pre-mixed resin/aggregate system (see section below, "Pre-mixed overlays").

Both approaches are aimed at providing a waterproof protective layer for the reinforced concrete (or steel) deck, in addition to giving a skid resistant and durable wearing surface. As discussed in Section 3.4, concrete decks may also be protected by polymer impregnation but the use of this technique in the UK has been limited.

Overlays built-up in-situ

It has been estimated that, by 1987, at least 100 bridges in the USA and Canada (including some new bridges) had been overlaid with "resin-concretes" (more correctly resin-mortars) based on epoxies, (meth)acrylates, and, predominantly, polyesters (Fowler, 1987). Such overlays have also been used in the USA for car parking decks (Mendis, 1987).

The majority of these overlays have been formed in several layers by initially spraying an unfilled resin onto the prepared surface, using truck-mounted metering and dispensing equipment. The resin is then brushed-in prior to broadcasting an aggregate, the excess being swept away once the resin has hardened. This process is repeated until three or four layers have been applied, giving a protective, anti-resistant wearing surface with a thickness of up to 10–15 mm.

These overlays have the following benefits:

- an excellent bond to the concrete
- thin sections which minimise dead-loads and eliminate the need to reconstruct approach slabs and kerbs
- excellent mechanical properties
- good skid resistance in wet conditions
- fast cure which minimises the bridge closure time.

Although durable cost-effective performance has been obtained from the early/mid-1970s onwards, many failures of earlier overlays were experienced. These have been attributed to the following:

- poor surface preparation resulting in loss of adhesion
- excessive interfacial stress resulting from mechanical movement, overly stiff/ brittle resins and the significant difference in thermal coefficient of expansion between the concrete and the overlay
- the use of too large and too angular aggregate which gives rise to capillaries and consequent suction of the binder from the surface by a "wicking" action. This resulted in poor wear resistance and inadequate protection against de-icing salts due to the formation of pinholes, often in clusters. This type of failure was particularly prevalent when only a single resin/aggregate layer was used.

Improved application procedures, more appropriate aggregates (in terms of particle size, shape and wear resistance) and the use of the "multi-layer" technique contributed significantly to the later successful performance of these overlays. The use of tough resin binders with a low modulus of elasticity and high elongation at break (often 20–50 per cent) has also been critical to the elimination of concrete failure close to the interface (due to differential thermal expansion/contraction). Overlays with a modulus in the range 1–8 GN/m^2 are now commonly employed. Good wear resistance is obtained and the useful life of ten years which can now be anticipated (Sprinkel, 1989) provides a cost-effective means of maintaining bridge-decks.

Most of the work on these overlays has been conducted in the USA and the developments there can be traced through many papers given at the ICPIC and ACI Conferences (see Bibliography).

Pre-mixed overlays

These overlays are formed directly from pre-mixed resin/aggregate (filler) systems. Protection against reinforcement corrosion is still a main function in addition to skid and wear resistance. Polymer (latex) – modified concretes and mortars have also been widely used for similar re-surfacing, particularly in the USA.

Trowel applied epoxy/calcined bauxite overlays were used on steel bridge decks in the UK during the 1960s (Gilmour, 1969) while a variety of systems were placed on removable steel carriageway panels during a trial which commenced in 1970 (Mander, 1981). Failures in this trial were attributed to insufficient flexibility for accommodation of traffic-induced deflections (particularly at low temperatures). The use of trowel-applied systems for concrete decks has also been described but no specific applications were noted (Mander, 1981).

Resin concrete overlays built-up in-situ (see above) have been used in the USA since approximately 1960. However, it was realised quite early on that improved performance was likely to be obtained from pre-mixed systems (McConnell, 1966). This approach could not be adopted until methods were available for rapidly mixing and placing large volumes of mortar having a short pot-life. The first use of pre-mixes appears to have taken place in about 1983.

The re-surfacing of concrete bridges on the Massachusetts Turnpike was carried out with a low shrinkage methacrylate-based material, placed at a thickness of about 100–150 mm, thus embedding the new upper reinforcement (Kane, 1984). A modified mobile concrete mixer was adapted for this work and produced 10–12 tonnes of polymer concrete within about 40 minutes. Traffic disruption was minimal as the work was carried out between peaks and the resin gave a good compressive strength (27 N/mm^2) within one hour. A relatively flexible system was used (compressive modulus of approximately 8 GN/m^2) to alleviate interfacial stresses due to differential thermal movement.

A related application on the Brooklyn Bridge utilised a relatively resin-rich epoxy slurry which was batched and dispensed using commercially available equipment more often used for pre-cast resin concrete. This slurry was screeded to a thickness of about 12 mm prior to broadcasting additional aggregate for skid resistance (Scarpinato, 1984).

A number of well-known bridges in the USA, for example San Mateo, Oakland Bay, Ben Franklin and Golden Gate, have been surfaced with an epoxy asphalt concrete which was developed in the mid-1950s. Some of these pre-mixed overlays, generally about 18 mm thick, were reported to be in excellent condition after 20 years service (Fowler, 1987).

An electrically conductive resin concrete overlay, based on a vinyl ester, has also been developed and applied to a bridge deck in Virginia, USA which had deteriorated due to chloride induced reinforcement corrosion (Fontana, 1989). This overlay has the dual function of providing a wearing surface and a means of distributing a low-level cathodic protection current uniformly across the deck surface.

It is also relatively thin (10–20 mm) thus giving a simple transition between the bridge and roadway; additionally, kerb-lines and the dead-load are essentially unaffected (disadvantages of the electrically conductive asphalt-concrete overlays developed earlier for the cathodic protection of bridge-decks).

3.4 RESIN INJECTION AND POLYMER IMPREGNATED CONCRETE

The use of a differential pressure for the controlled filling of fine cracks and small voids is one of the earliest applications of resins in construction, dating from the late 1950s. It is also one that cannot be achieved with other types of material, particularly where there is a structural demand.

This application, discussed in Section 3.4.1, is distinguished from the grouting of larger voids and gaps (often of a pre-set size – see Sections 3.7 and 3.9) by the common use of unfilled resin systems. Although this may suggest an application bearing a high material's cost, the volume of resin required is usually extremely small. Labour costs may be high as effective work often requires both patience and skill. Nevertheless, considerable financial benefits have frequently been obtained by using resin injection to re-instate pipes, beams, columns, walls, floors, dams, piers, piles and tanks.

The improvement of concrete strength, permeability and durability by polymer impregnation was considered, in the 1960s/1970s, to have enormous technical and economic potential, particularly in the USA and Japan. This topic is discussed below (Section 3.4.2) as, in common with resin injection, unfilled polymers are employed beyond the immediate surface zone of the base material.

This depth of penetration distinguishes the application from the surface sealing discussed in Section 3.2. However, polymer impregnation is treated relatively briefly as its potential has not been truly fulfilled, despite major research and development effort.

3.4.1 Resin injection

Scope

One of the first applications of resin injection took place in 1959 during attempts to repair a concrete dam in California (Boyes, 1985). Since that time, the technique has evolved to the point where it can be used routinely by specialist applicators to meet various demands.

Some examples are:

- monolithic reinstatement of cracked reinforced concrete, stone or brick to its original design capability
- fine crack sealing and small void filling in concrete and masonry (for preventing leakage or the penetration of water or other corrosive materials and/or structural consolidation)
- re-fixing partially de-bonded tiles or renders
- re-bonding delaminated floor toppings
- the reinstatement of split or delaminated timber
- the in-situ formation of polymer concrete (by injecting pre-placed aggregate, eg in large, accessible voids or gaps; this approach can be very useful for underwater patch repair).

Resin injection can be carried out in wet conditions and there is considerable world-wide experience of successful use underwater, in some cases at appreciable depth. For this work, the resin must be capable of displacing the free water (while maintaining a stable boundary) and absorbing that on the substrate surface (while still curing effectively). Both chemical and physical properties such as viscosity and density must, therefore, be appropriately controlled.

When assessing various resins for crack filling, it is essential to distinguish between the internal (real) crack width and that which is visible at the surface (where widening may have occurred due to minor spalling) – see Figure 3.8.

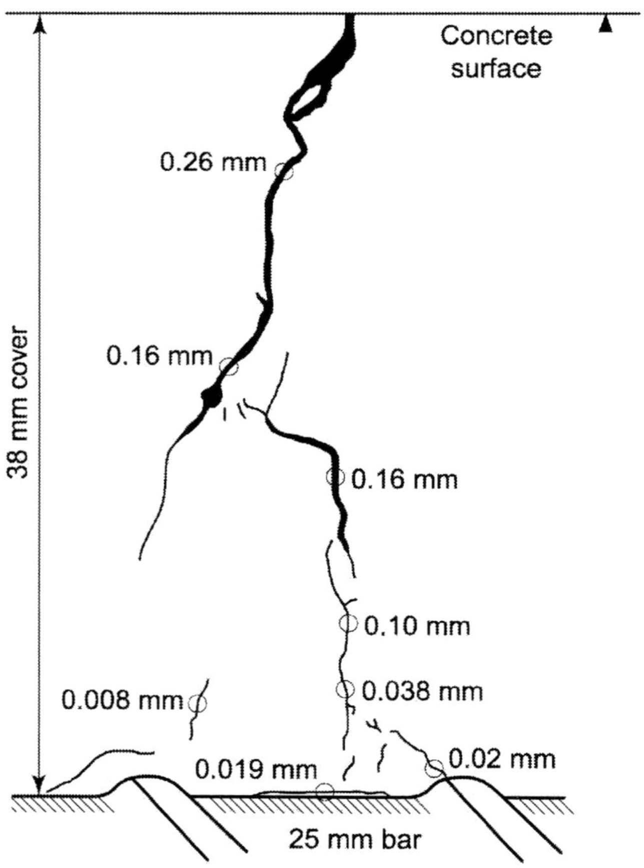

Figure 3.8 *Variation of crack width with depth (The Concrete Society, 1992)*

With an appropriate combination of resin formulation, equipment and operator skills, cracks which are approximately 0.2 mm wide at the concrete surface can be injected. Such cracks often taper, but good penetration into a width of less than 0.05 mm can be achieved.

Where extensive cracks have to be filled completely (to meet a structural, adhesive or sealing requirement) experience and a carefully planned approach are particularly important; partial filling will usually be impossible to remedy at a later date. The success of resin injection may often be assessed by core removal for visual inspection/bond testing or, in some instances, by partial coring to give an in-situ test piece containing a bond-line which is subjected to a direct pull-off stress.

As the feasibility and effectiveness of resin injection (and the time required for the work) are frequently very difficult to assess in advance, pre-contract trials may have to be considered.

Any possibility that a low viscosity resin will drain from an injected crack prior to gelation has to be addressed. In very fine cracks, surface tension (capillary) forces may assist retention of the liquid resin. However, with wider cracks, it is often necessary to employ a thixotropic resin. Provided the degree of thixotropy is selected carefully, penetration down to a width of 0.05 mm can still be achieved. Such resins are very useful where any remote crack opening cannot be surface sealed, for example, due to lack of access.

At crack or gap widths of about 5–10 mm and beyond, the use of filled resin or cementitious grouts would usually be preferred to neat resin injection on cost grounds. Additionally, an unacceptable exotherm could occur with an unfilled resin; this could lead to thermal contraction stresses or even boiling and charring of the resin.

Pre-assessment and limitations

It is imperative and cannot be emphasised too strongly that the following facts in particular must be established prior to treating cracks by resin injection:

- **the reason(s) for cracking**
- **whether the cracks are stable, progressing or undergoing significant contraction and expansion.**

Three examples of resin injection being used inappropriately are given below.

1. The injection of cracks, in concrete or masonry for example, which are caused by settlement or continuous overloading (or under-design). Unless some permanent form of support or stress relief is implemented, progressive strain after resin injection is likely to cause adjacent cracking.
2. The injection of cracks caused by restraint or a lack of movement (expansion) joints. The injection of such active cracks is often considered because resins have a lower modulus than concrete and very good adhesion can be achieved, suggesting that the movement can be tolerated. However, for very many resins used for injection (eg, epoxies), the modulus is likely to be at least several orders of magnitude higher than that of a typical joint sealant (<5 MN/m^2), particularly at low temperatures. Even if a crack is filled with a very low modulus resin, sufficient stress can still be generated to fracture the adjacent concrete substrate (see Figure 3.9(b)). For example, the opening of a 0.5 mm wide crack by 2.5 microns can rupture the concrete (Keer and Emberson, 1986). To accommodate movement in service, conventional expansion joints have to be designed with a width above a

certain minimum figure, thereby limiting strain (and hence stress) to an acceptable level (see Appendix A1.4). Cracks due to the omission of expansion joints can be injected provided a relief joint is formed in an adjacent position – shown as (d) in Figure 3.9, where other options are also given.

3. The injection of cracks in concrete caused by reinforcement corrosion (for long-term repair). These usually coincide in position and direction with the corroding steel, may be associated with rust staining at the surface and often precede extensive delamination of the cover. While delamination may be delayed following injection, continued corrosion will generate adjacent cracking in due course (in most cases, significant inhibition of further corrosion, by resin "encapsulation" of the steel, is unlikely to be achieved). Consequently, injection can be considered only as a temporary measure.

Resin injection can be used to rectify poorly made construction joints and to restore structural integrity where cracking has been caused by temporary overload; in the latter case, cracks in concrete will commonly occur diagonally or transversely to the direction of the main reinforcement due to shear or tensile/flexural stresses. Cracks due to the absence of properly formed contraction joints (which accommodate cure shrinkage) may also be injected once they have reached a maximum width

If the cracks have existed for some time in these or other cases, significant reinforcement corrosion, carbonation or chloride ingress may have occurred and should be assessed prior to resin injection.

The existence of cracks does not necessarily imply a need for repair. Most reinforced concrete members are designed so that limited and controlled cracking is expected under normal conditions. Such intrinsic cracking must, therefore, be distinguished from that due to other factors. The direction of cracking, particularly in relation to the reinforcement, and the time of its occurrence are informative in this respect (The Concrete Society, 1992).

In addition to the reason for cracking, crack width and the exposure conditions also contribute to the assessment of cracked reinforced concrete. More detail needs to be considered to determine whether any action is necessary and then to define an acceptable strategy for repair; the reader is referred, therefore, to the Bibliography. However, where injection is an appropriate option for sealing and/or structural reinstatement , then the technique provides a powerful and versatile approach when used by an experienced contractor.

The response of reinforced concrete members repaired by resin injection to uni-directional (static) and cyclic stress has been investigated in some detail (eg Hewlett and Morgan, 1982; Keer and Emberson, 1986; Chung 1975 and 1981). This laboratory work dealt with initial failure due to both flexural tension and shear. It was established that, if the cracks are accessible to the resin, failed specimens can be re-instated to load-deflection and ultimate load behaviour at least as good as that exhibited prior to failure. As flexural and shear cracks are mainly continuous, passage of the resin tends to be limited only by very small crack widths.

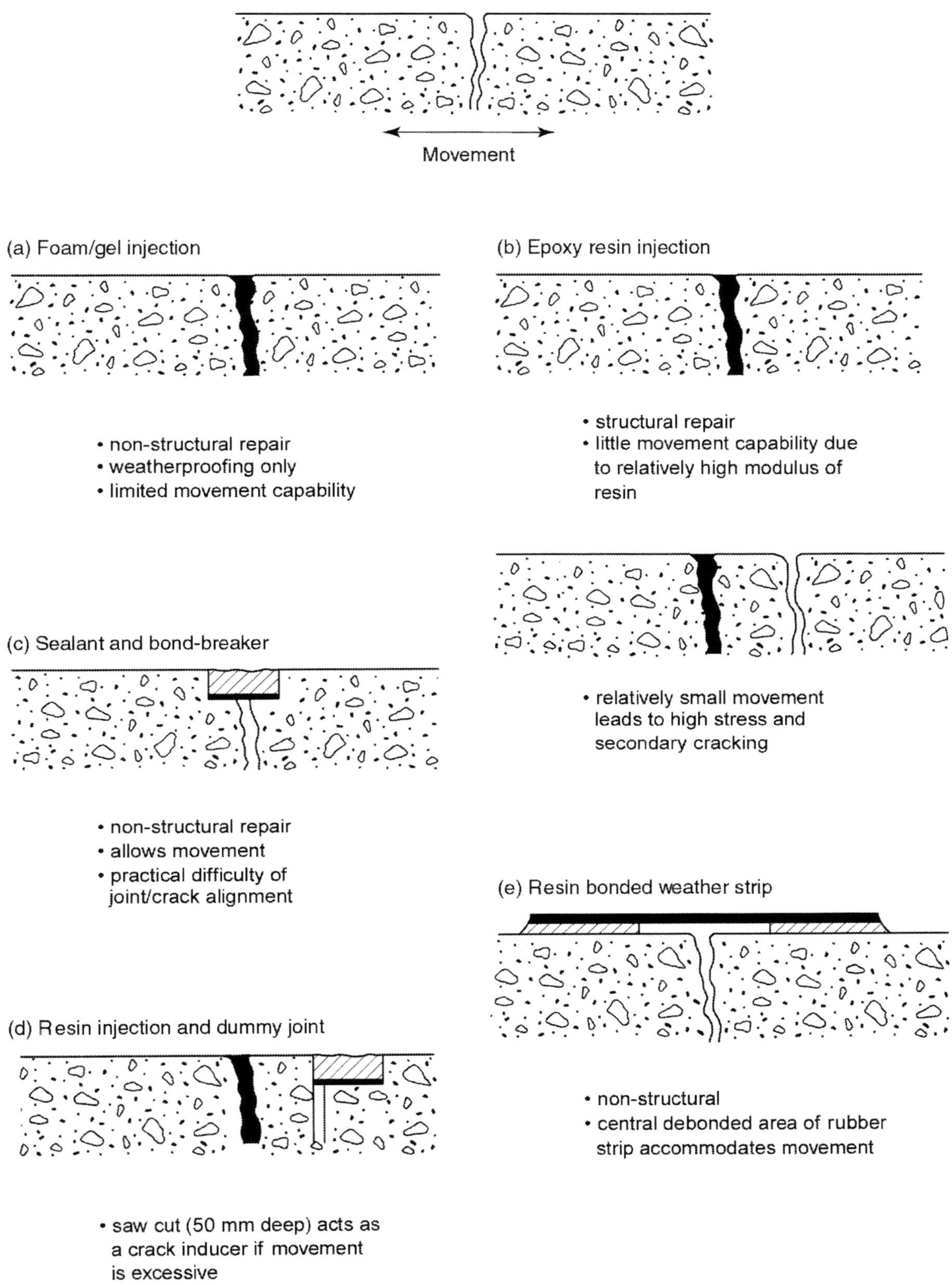

Figure 3.9 *The treatment of cracks in concrete*

If there is bond failure between the reinforcement and concrete, total reinstatement is unlikely as the longitudinal cracks along the bars are usually discontinuous and narrow and, therefore, difficult to inject completely (even though they may extend to the concrete surface). However, sufficient re-bonding, via injection, may be achieved for an effective repair (Chung, 1981). Similarly, concrete containing ribbed bars can be repaired successfully if the secondary cracking associated with the ribs can be injected.

The securing of partially de-bonded external tiles and rendering by resin injection needs to be approached with due care as it can be extremely difficult to ensure that existing gaps are completely treated. If water becomes trapped in any remaining gaps, freeze-thaw damage may result in due course. Slow water uptake by the resin itself also leads (potentially) to slight expansion – however, relaxation effects are likely to offset the tendency for any significant stress to develop.

To avoid further tile or render de-bonding when performing this work, injection pressures must be strictly limited. Extreme care in the use of very reactive resins is also necessary as exothermic expansion may lead to damage. Consequently, if large voids are associated with de-bonded tiles/renders, low activity resins or other materials should be considered.

Differential thermal expansion/contraction between the resin and substrates is unlikely to cause problems where the bond-line is thin.

Resin type

Epoxy resins are very commonly used for injection work, particularly where service stress transfer is required as, with appropriate formulation, they have the following attributes:

- suitably low viscosities are readily achievable, thus assisting rapid injection at low pressures; where required, thixotropic products can be formulated
- cure shrinkage is minimal (both before and, more importantly, after hardening)
- they are suited to use with proportioning and dispensing equipment (this benefit is discussed further below)
- a wide range of pot-lives and cure rates can be provided, with allowance for different application temperatures
- excellent adhesion can be obtained (and maintained) even in very wet or underwater conditions and on alkaline concrete surfaces
- a high modulus can be obtained (for efficient stress transfer).

The main limitation of most epoxy resins for this work is their limited ability to cure at very low temperatures (<5°C).

It is essential that solvents are not added to these (or any other type of resin) to assist penetration. Solvent evaporation will lead to detrimental shrinkage; entrapment of small concentrations may also occur, causing poor resin cure and adhesion.

Polyester and (meth)acrylate resins are employed for injection work although, for many cracks, they are limited by cure shrinkage and relatively poor adhesion in wet conditions. However, they can be used for void filling and consolidation (eg old masonry) and where service stresses are mainly compressive, eg re-bonding delaminated floor toppings. They also provide good low temperature cure and lower viscosities than many epoxies, but batch mixing rather than continuous metering, mixing and injection is usually necessary.

Polyurethanes, although little used in the UK, are employed for some proprietary injection processes.

In certain circumstances, materials other than resins can be used for crack filling, for example: cementitious slurries and grouts (possibly polymer modified), polymer latex dispersions, chemical grouts and injectable polymer systems which gel or foam in-situ. The latter, often based on polyurethanes or very similar materials, may allow foaming on contact with water to be controlled.

Procedures

The resin injection procedure may be divided into the following phases:

- pre-treatment of the cracks
- surface sealing with appropriate spacing of injection/air release points
- proportioning, mixing and injection of the resin
- removing the injection ports and surface sealing material to leave a satisfactory appearance.

Pre-injection preparation

The pre-treatment of cracks is not required routinely and is only likely to be considered when contamination or excessive dust/debris is present. The latter may be removed by vacuum or dry, oil-free compressed air but flushing may be necessary to remove contaminants such as oil. This is invariably a difficult operation of uncertain efficacy and one that is potentially hazardous if solvents are used; solutions of detergents or other surfactants may be used provided they do not subsequently affect resin adhesion.

An appropriate positioning of injection/air release points has to be determined for each case and needs to take into account the depth of penetration required. For many cracks, it may be assumed that the resin flows radially and uniformly and, therefore, the spacing of these points needs to exceed the crack depth to ensure complete penetration. This approach should be tempered with experience as the assumption is over-simplified where there are obstructions or significant variations in crack width.

When injecting delaminated floor toppings, injection points are often set close to the limits of hollowness, rather than along the line of any associated cracking, to avoid trapping air.

Various materials are used to provide a surface seal between injection points, for example: fast curing pastes or mortars (resin-based or cementitious) hot melt thermoplastics and special tapes. Selection will be governed by such factors as crack width, likely injection pressures, ease of removal and the final appearance desired (or specified); some materials are initially pigmented to colour match the surface.

A number of methods are also available for providing resin entry points, for example plastic tubes set into drilled holes, surface bonding of metal tubes with splayed ends and "self-sealing" rubber nipples on the "injection gun". Where holes are drilled into a fine crack, the use of hollow masonry bits and a vacuum attachment should be used to prevent blocking with fine debris; allowance must also be made for cracks which do not travel into the concrete at right angles to the surface. Continuity between the openings provided for resin entry/air release can be checked by air flow, indicated by a simple bubbler.

Injection

Three distinct injection techniques, each with a number of detailed variations, are widely used.

1. Batch mixing of the resin which is then injected under a positive pressure.
2. Continuous mixing and injection under pressure using self-contained and purpose-built equipment (see Figure 3.10). When injection requires a resin of paste consistency, partitioned cartridges containing both resin and curing agent/hardener (see Volume 3) and fitted with a static mixing nozzle may be used, ie a simple version of the continuous mixing/injection technique.
3. A process whereby penetration is vacuum assisted.

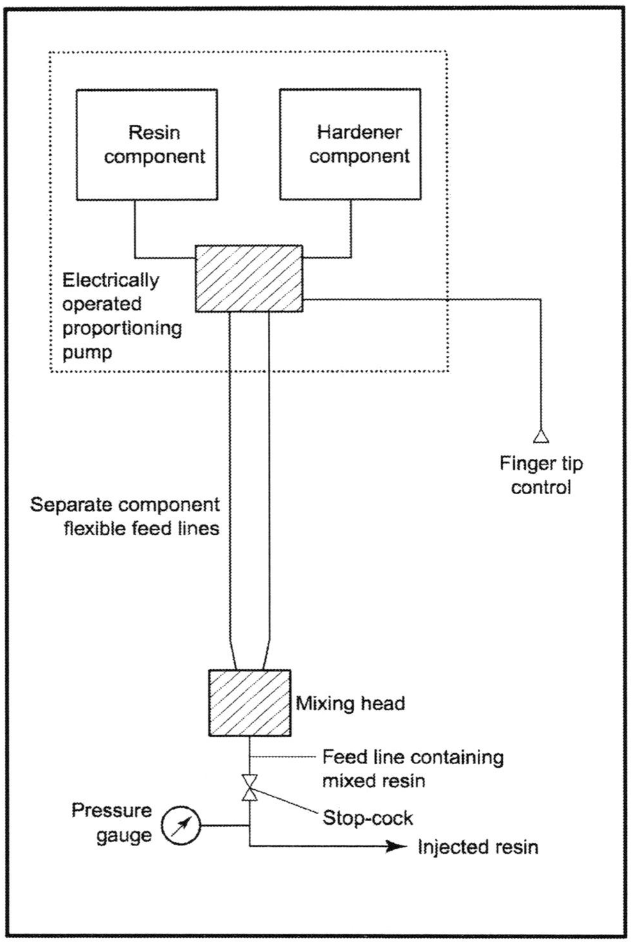

Figure 3.10 *Resin injection equipment*

The injection of batch mixed resins can be carried out by pump, sealant gun or, less often in recent work, by use of a pressure-pot. With some proprietary resin injection "kits", the use of a sealant gun is facilitated by the provision of standard plastic cartridges in which the resin can be mixed. Semi-automatic injection of a batch mixed resin can also be carried out using a patented system which utilises rubber "balloons" containing resin under pressure (see Figure 3.11).

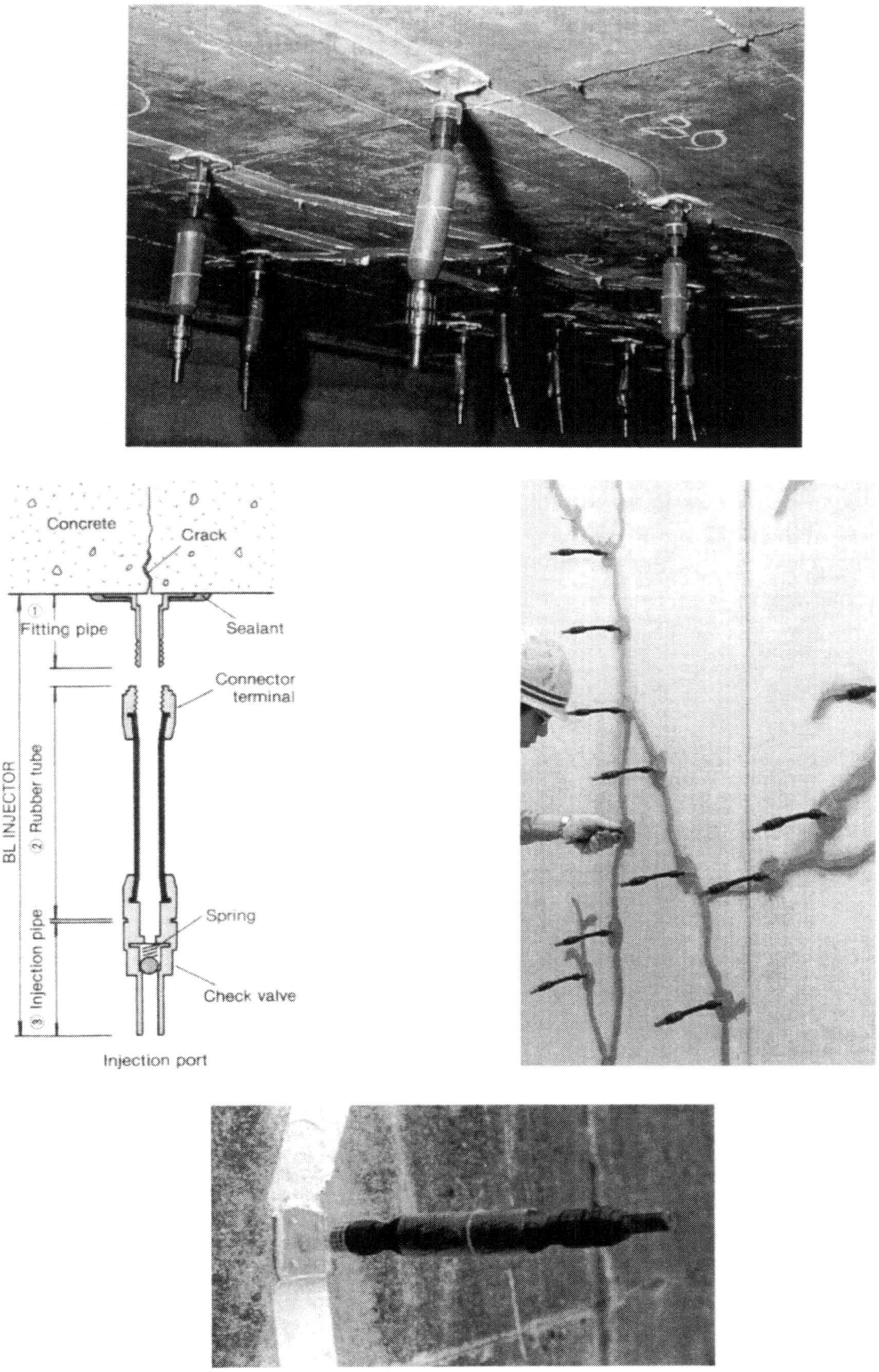

Figure 3.11 *Semi-automatic resin injection system*

Batch-mixed injection

Provided fairly wide cracks are being treated, gravity feed using a simple fluid head may be feasible. In practice, however, this technique frequently causes problems and is probably better avoided. Injection with batch mixed resins can be satisfactory where penetration is achieved easily. For more difficult work, it can become impractical because the injection time of the mixed resin is limited by its pot-life – an even earlier loss of efficiency can occur with any progressive and significant increase in viscosity. This problem is exacerbated where it is difficult to mix the resin close to the point of injection eg on scaffolding, cradles or underwater.

Proprietary "injection kits" which rely on the use of sealant guns can be very useful provided that certain requirements and limitations are acknowledged, for example:

- they should not be used by the general building contractor as "a simple maintenance procedure". A proper assessment of whether resin injection is actually suitable should be made by an experienced specialist prior to use
- secondary damage may be caused by the uncontrolled and unmonitored generation of high injection pressures, particularly when de-bonded tiles or renders are injected; when treating cracks in concrete, the surface seals are likely to be forced off, resulting in partial crack filling
- partial injection resulting from any other unforeseen practical difficulties can be virtually impossible to correct at a later stage, even by a more experienced contractor: essential flow paths may be blocked; drainage will leave a resin coating on the substrate, resulting in poor adhesion on re-injection. It is essential "to get it right first time".

Continuous injection

Self-contained equipment which has been specifically designed for the continuous metering and mixing of components and injection provides the following benefits:

- the equipment tends to confine specialist work to appropriately skilled and experienced contractors
- exposure to the resin, equipment cleaning and contaminated packaging is minimised
- the formulated components can be purchased in bulk (rather than as relatively small pre-weighed packs)
- very long component supply lines can be fitted between the proportioning pumps and the resin-mixing head. This can greatly simplify working under water or where space is limited
- very small quantities of resin are mixed at any time, allowing slow injection to be carried out continuously even with fast reacting products. As the gel time often increases with a decrease in resin volume (see Volume 3) the time for which a low viscosity is maintained can vary usefully in the following order: fine crack > small mixing head > large batch from pre-weighed packs. Additionally, resin in the mixing head can easily be discharged to waste with virtually no interruption to a slow injection process
- the process can be controlled completely and rapidly by fitting a pressure gauge to the hand-held mixing/injection head and providing the operative with a spring-loaded switch for the supply pumps. Thus, secondary damage due to excessive pressure can be avoided when injecting de-bonded tiles, renders or floor toppings. Additionally, a fluctuating injection pressure can easily be employed – this is more effective than a continuously applied high pressure which forces the resin along the path of least resistance, leaving voids.

Typical equipment used for resin proportioning and injection is shown in Figure 3.12. In general, this equipment benefits from a simple means of adjusting the mixing ratio, thus allowing a range of resin systems to be used; precise proportioning under a variety of ambient conditions must also be guaranteed (this can easily be checked by providing appropriate back-pressure and measuring the delivery from each supply line). A means of heating or cooling the resin components can often be incorporated with relative ease and, on some equipment, the following facilities are provided: adjustable transducers which automatically limit the injection pressure; optical sensors which switch off the pumps if the supply of either component is interrupted; shear pins which provide a fixed upper limit to the operating pressure; automatic metering of the dispensed volume; control of the pump speed.

Vacuum-assisted injection

When a positive injection pressure is used, air (or water) must be readily displaced from the crack or void as trapped pockets can prevent full penetration. For a "simple" vertical crack, this may require no more than the "standard" procedure, ie commencing injection at the lowest entry port and then moving to the next one above only after resin has travelled slightly beyond this point. However, where a more complex network of cracks exists and incomplete penetration due to trapped air is likely, a process involving application of a vacuum prior to injection is preferred. This technique is particularly applicable to isolated objects, such as statues, and large flat areas containing a visibly extensive network of cracks, for example, fissured or crumbling brickwork, stone, rock or concrete. Where the substrate is also porous in a more general sense, a degree of resin impregnation contributes to the improvement achieved.

This method, which has been in use for at least 20 years, requires initial shrouding of the object or area to be treated with a flexible and airtight membrane, normally polyethylene sheet. A vacuum is applied beneath this membrane, resin is introduced via appropriate ports and the vacuum is maintained until gelation has occurred. Resin is thus sucked into the cracks and voids under atmospheric pressure. Although beneficial in avoiding secondary damage during injection, this relatively low pressure restricts the vertical flow of resin; in some circumstances, therefore, a staged process may have to be adopted.

Limited access for positioning of the sealing membrane and remote air pathways which cause a loss of vacuum also present practical difficulties, although these can usually be overcome.

Compared to injection under an applied pressure, the vacuum-assisted process has the following benefits (in addition to those mentioned above):

- moisture can be removed easily under vacuum
- air removal assists resin penetration at the apex of cracks, ie the point of potential propagation
- adhesion may be improved due to greater resin/substrate contact within the "troughs" on rough surfaces
- the existence of leakage paths may be revealed by difficulties in achieving a good vacuum.

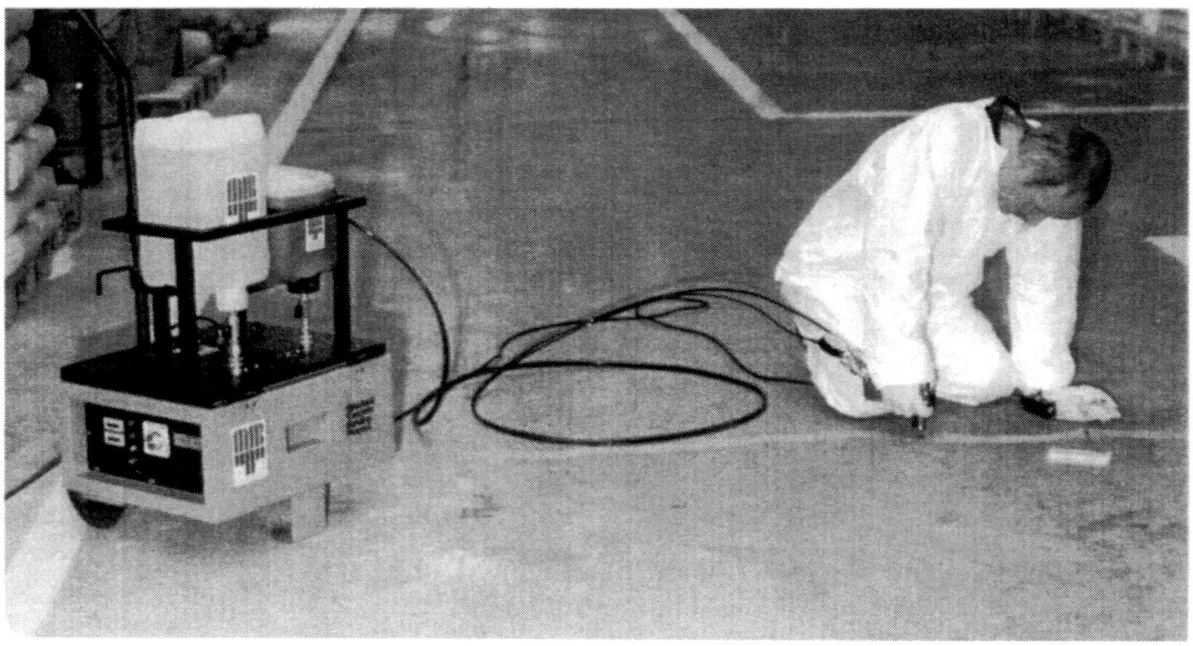

Figure 3.12 *Resin injection unit*

The inevitable formation of a resin-coated surface can be a disadvantage in some circumstances, eg for aesthetic reasons.

Epoxies, polyesters and (meth)acrylates are used with this technique. In certain circumstances, other materials are used, for example, silanes, silicones, cementitious and other special grouts. The following applications have been reported (Heayes, 1979; Ridout, 1982; Acton, 1979; Bond, 1980; Anon, 1980; Guthrie and Milne, 1983; Hayward, 1982; Anon, 1982):

- sealing of cracked concrete car parking decks and nuclear storage ponds
- consolidation (and weatherproofing) of porous, cracked brick and stonework, eg walls, panels, statues; load-carrying concrete bases, eg beneath crane-rails; bridge-bearing support pads
- remedial grouting of post-tensioning ducts, eg concrete bridge beams
- voids grouting beneath rocking or subsided slabs, eg roads, airports, factories; a relatively light and adaptable space-frame is used to raise subsided slabs prior to injection.

Sealing of construction joints

The injection techniques discussed above are only considered when a problem becomes apparent, ie post-construction. However, one additional process, which has been extensively used in Europe, Japan and the USA for some years and is now available in the UK, anticipates the future need for injection at the construction stage. This system provides a specific and simple solution for the sealing of construction joints in concrete, ie a very common source of water leakage.

The specially designed and patented flexible hose (illustrated in Figure 3.13) is clipped along the full length of the hardened concrete joint face. This hose, which is fitted with purpose made end-pieces for injection and venting, can be used to water test the joint and then to carry out injection should sealing be necessary. This system is presently used in the UK, with either a special cementitious grout or with a very low viscosity resin-based product. The latter is an acrylic form of vinyl ester which gives a tough, flexible and hydrophilic material with the property of becoming swollen with water.

A related system originating in Germany is also available in the UK. This system incorporates specially designed hose connectors which are fixed to the shuttering. Injection, where necessary, is carried out using a two-pack polyurethane product.

Summary

Due to its versatility and potential benefits, particularly for remedial work where there is no other viable option, resin injection is now a widely employed technique. However, due regard must be paid to its limitations and practical difficulties. A summary of key factors associated with the technique is given in Table 3.4.

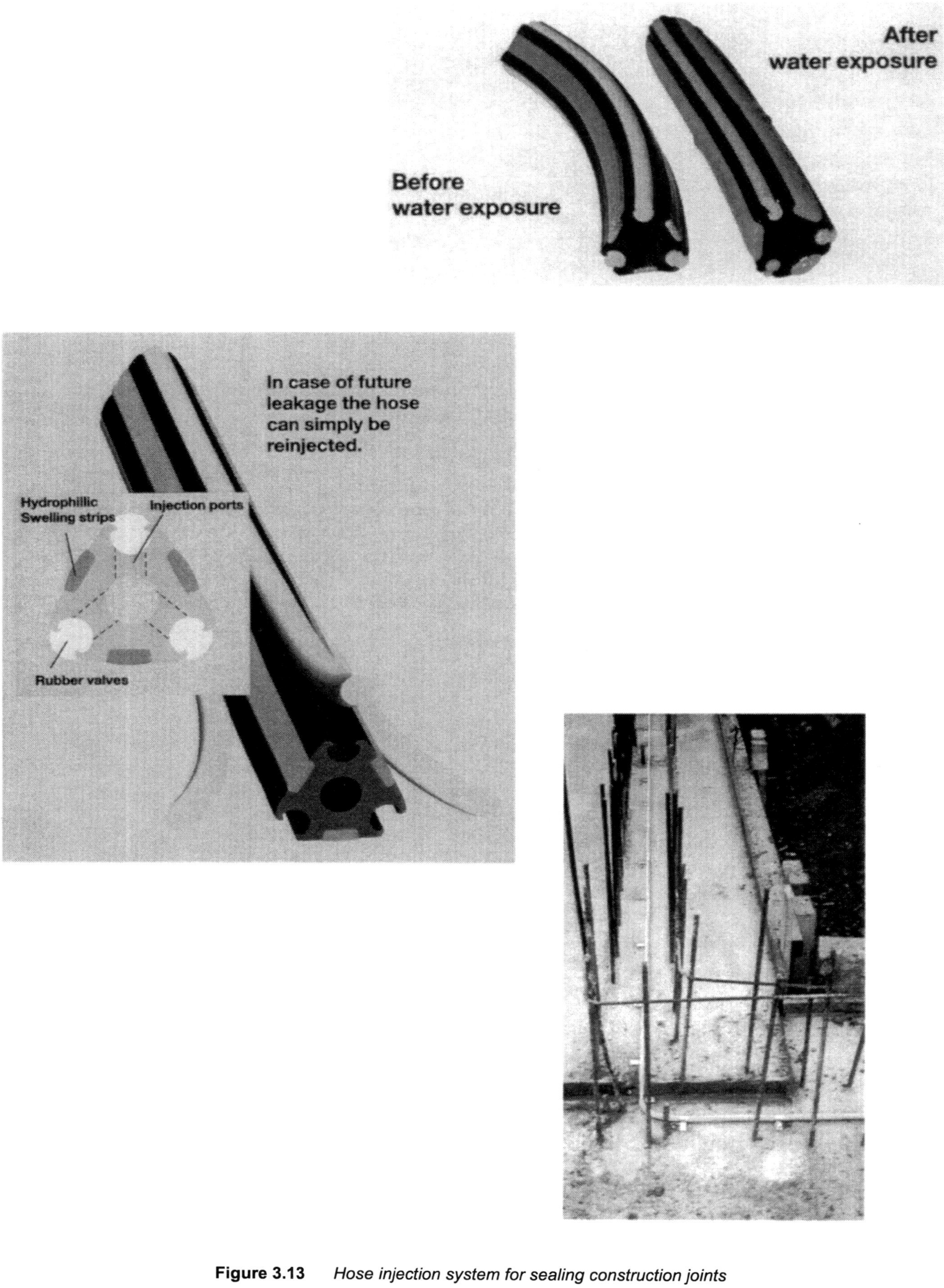

Figure 3.13 *Hose injection system for sealing construction joints*

Table 3.4 *Resin injection – key factors*

Do	Do not
Determine the reason for cracking and assess whether injection is suitable for reinstatement	*Use injection to repair active cracks caused by a lack of expansion joints, reinforcement corrosion (other than as a short-term measure) or overload (until stress relief has been implemented).*
Determine whether sealing is required for subsequent durability, the reinstatement of structural integrity, the prevention of water ingress or only appearance.	*Ignore other (and possibly more simple) options, including "no action necessary".*
Use specialist and experienced contractors and consider the need for pre-contract trials.	*Treat injection as a simple maintenance procedure for the general contractor.*
Check crack continuity and consider the best means of forming access ports.	*Block injection paths with dust from drilling.*
Consider whether leakage paths are likely to be present (including existing expansion joints which injected resins should not be allowed to enter).	*Use a very low viscosity resin if a thixotropic system is more appropriate.*
Consider the likely internal crack widths and the need for full depth injection (see Figure 3.8).	*Use excessive pressure for complete crack filling, to reduce the injection time or to compensate for high resin viscosity at low temperatures.*
Determine the most suitable positioning and spacing of access ports.	*Ignore the need for air or water flow from the crack or void.*
Ensure that the resin is suitable (viscosity, pot-life, shrinkage, adhesion, water displacement) and allow for the effects of ambient temperature.	*Use fast curing exothermic resins in wide cracks or large voids; or dilute any resin with solvent.*
Select an injection technique (and resin) which is proven and suited to the specific task.	*Attempt to use batch mixing and a gravity feed for cost saving or to avoid employment of a specialist contractor.*
Ensure that resin systems are always proportioned and mixed correctly.	*Ignore the need to check equipment frequently and thoroughly, particularly when working at temperature extremes.*

3.4.2 Polymer impregnated concrete

Scope

This application is predominantly concerned with the impregnation of relatively good quality Portland cement concrete using a monomer (rather than a resin) which is subsequently polymerised in situ (ACI, 1986). Effectively, this gives a new composite material with markedly different properties to normal concrete, particularly when full depth impregnation is achieved. The process may be viewed as the ultimate stage in a sequence which includes the resin injection of cracks and voids and the sealing of unsatisfactory surfaces, as discussed in Sections 3.2.2 and 3.4.1.

Properties

The primary effect of impregnation is to modify the internal structure of the cement phase by sealing the continuous capillary pore system; stress concentrations arising from the presence of pores and micro-cracks are thus reduced. Modification of the cement/aggregate bond is usually considered to be of less significance.

These changes lead to a marked improvement in mechanical properties, permeability and thus durability:

- strength can be increased by a factor of at least four and modulus by two
- creep can be considerably reduced
- water absorption can be reduced by 99 per cent with a major improvement in freeze-thaw resistance; chemical resistance is also increased markedly.

When high strength is of primary concern, full impregnation is essential. This is only feasible for pre-cast components which are factory impregnated as the concrete must be very dry (150°C is often used) and fully saturated with monomer. Initial evacuation followed by complete immersion in the monomer (under pressure) is common.

For improved durability, surface or partial depth impregnation (eg 10–40 mm) is adequate and has cost benefits. Consideration can then be given to on-site impregnation of large surface areas, using a simple "soaking" method rather than the vacuum-pressure technique.

Polymer types

The most distinctive feature of polymer-impregnated concrete is the use of monomers rather than more conventional resin systems. These monomers, which may include low molecular weight cross-linking agents, have an extremely low viscosity with obvious benefit for deep penetration. Vinyl monomers, either alone or as blends, have been most commonly used (or assessed) for example, various acrylates/methacrylates, styrene and acrylonitrile. Silane coupling agents are frequently incorporated.

Polymerisation occurs in a similar manner to the cross-linking reaction associated with unsaturated polyester resins and can be induced by a similar combination of catalyst and accelerator. However, this method can limit the time available for saturation and, consequently, a thermal-catalytic technique is frequently employed; in this case, heat is used to "activate" the catalyst rather than a chemical accelerator. This allows the onset of polymerisation to be controlled while also providing a short processing time. Radiation induced polymerisation has also been investigated (and used in the past) but it is an expensive method and has a number of practical difficulties.

The monomers are, in general, volatile, flammable and, to varying degrees, toxic. Consequently, they present health and safety hazards and, also, problems due to evaporation loss during application. Appropriate measures are essential for good storage stability.

A more significant problem in relation to the relative lack of widespread use is the high cost of these monomers. As a result, considerable effort has been given to maximising the improvement in performance with a minimum consumption of monomer.

Epoxy resins have also been investigated but, although they can provide excellent performance, high viscosity limits their use. Where only partial impregnation is adequate, ie of the surface, polyester and (meth)acrylate resins have also been used.

Applications

The potential areas of application for fully impregnated concrete include pre-cast tunnel linings, pipes, reinforced/post-tensioned beams, and various components used in marine and other aggressive environments. Partial impregnation has been aimed mainly at the protection of concrete bridge decks and spillways.

Despite its potential benefits, a combination of cost factors and technical difficulties has limited the use of polymer-impregnated concrete (eg doubts concerning adequate crack sealing on bridge decks exposed to chlorides and, also, the need for dry concrete). However, considerable detailed research has been carried out on all aspects of this material, particularly in the USA and Japan. The progress of this work can be followed in the proceedings of the ACI and ICPIC conferences (see Bibliography).

3.5 SEGMENTAL CONSTRUCTION

Introduction

Virtually all on-site applications of resins in construction depend on achieving and maintaining good adhesion. The use of resins as adhesives requires similar or dissimilar surfaces to be joined so that bonded components act as a composite unit. Such uses may be truly structural, in the sense that the adhesive resists stresses that would otherwise lead to separation, or they may be semi-structural, ie compression is a predominant component of the applied stress. In either case, thermosetting resins provide many benefits, for example:

- a fast cure rate under many site conditions
- high mechanical strength and an adequate resistance to creep
- inherently good adhesion to a wide variety of surfaces
- maintenance of performance under a wide range of service conditions(although upper service temperatures may be limited)
- availability in different physical forms and as thixotropic products
- good gap-filling properties without the need to apply pressure during cure (due to the absence of volatiles such as solvents or reaction by-product and a relatively low cure shrinkage).

Examples of structural adhesive uses are discussed in Sections 3.6, 3.7 and 3.8.

General benefits

The present section deals with a semi-structural application where resins, invariably epoxies, provide a uniform stress-distributing (and waterproofing) joint between pre-cast reinforced concrete units which are post-tensioned.

The latter technique, conventionally known as segmental construction, has been used since the early days of pre-stressing (either with cast-in-situ, or, since the 1940s, pre-cast segments (Sriskandan, 1989)). In the particular case of bridges, it speeds construction and reduces costs as pre-casting can take place while the foundations are being formed. The incremental launching and pre-stressing of these units (as cantilevers from the piers) then avoids or minimises the need for falsework supports in the spans – an important benefit over rivers, estuaries or other areas with difficult access. For any structure, construction also benefits from pre-casting which can be carried out under "factory conditions" with a high level of control.

Until the early 1960s, joints were formed from cast-in-situ Portland cement concrete/mortar (or, for horizontal joints in columns, dry-packed mortar) and had widths ranging from 10 mm to over 250 mm. These joints posed a significant problem as pre-stressing could not take place until the concrete or mortar had cured sufficiently, an inherently slow process. Dry, non-bonded joints can also be (and are) used but this requires special keying for the resistance of shear stresses and very accurate mating for the prevention of water ingress and the uniform transmission of the pre-stress.

Examples

In 1963, for a bridge over the River Seine at Choisy-le-Roi, near Paris, the concrete segments were "match-cast" for the first time and a very thin layer (1–1.5 mm) of epoxy resin was then applied to both surfaces immediately before joining the units (Mays and Hutchinson, 1992). The rapid strength gain of the resin (less than one day) allowed the pre-stress to be applied with a minimal delay and construction could thus proceed almost continuously.

This adhesive use of an epoxy, although only semi-structural due to the compressive post-tensioning (and the frequent use of "shear/location keys" on the pre-cast units) represented a new and, particularly at the time, a technically bold application of resin systems. Since 1963, the impressive increase in the use of segmental construction, which may be attributed largely to the proven benefits of epoxy resins, has included many long-span bridges and other structures which are particularly prestigious and striking, both visually and technically. A number of these are noted in Table 3.5.

Particularly for structures such as Sydney Opera House, the Durban sugar terminal, and the Parc des Princes and Montreal Stadiums, adhesively bonded segmental construction allows complex shapes to be fabricated from much simpler, smaller units, thus facilitating the pre-casting and handling operations. Furthermore, critical geometry can be carefully checked prior to final assembly. Thus, changes of direction for horizontal or vertical curvature can be accommodated easily and reliably.

Construction procedure

Generally, the thixotropic resin paste/mortar has been manually spread over the concrete surfaces. However, on the Marron Bridge resin was pumped into 6 mm wide joints and on parts of the Montreal Stadium super-structure the wider (12 cm) joints were filled with a pourable resin grout (see Bibliography: Ciba-Geigy).

Table 3.5 *Some notable examples of resin-bonded segmental construction*

Structure	Comments
1960–1970	
Pont D'Oleron, Bay of Biscay, France	2900 m long bridge with, mainly, 80 m spans; constructed in 14 months
Pont Aval (and Amont), River Seine, Paris	34.6 m wide bridge on the Paris ringroad
River Volga bridge at Yaroslavl, Russia	Length 800–900 m
Casteljon Bridge, Pamplona, Spain	108 m length constructed from 208 segments and using 2 tonnes of adhesive
Rawcliffe Bridge, Dutch River, Yorkshire	The first UK use of segmental bridge construction (Dunster and Hough, 1968; Sims and Woodhead, 1968)
Ooster Schelde Bridge, Holland	5022 m length constructed in 2½ years
Chillon-Villeneuve Viaduct, Lake Geneva, Switzerland	1½ mile length
Sugar Storage Terminal, Durban, South Africa	Construction of the legs of 80 primary arches forming an 800 ft long structure
Sydney Opera House, Australia	Construction of the super-structure
1970–1980	
Park des Princes Stadium, Paris	Construction of 50 super-structure arches
Montreal Olympic Stadium and Velodrome	Construction of the super-structure involving approx 1200 thin (1.5 mm) joints using 30 tonnes of resin and approx 70 wide joints (up to 12 cm) using 100 tonnes of pourable epoxy grout
President Costa & Silva Bridge between Rio de Janeiro and Niteroi, South America	14 km long bridge opened in 1974 (then the 5th largest bridge in the world). 3224 segments bonded in less than 18 months with spans up to 300 m. Up to 12 segments, (up to 110 tonnes each) were positioned and bonded per day
Byker Viaduct, Newcastle-upon-Tyne, UK	815 m length
Marron Bridge, Cumbria, UK	172 m length; 6 mm wide joints formed by resin pumping
M180 Trent Bridge, Humberside, UK	Approx 364 units counter-cast off-site
Pasco/Kennewick Bridge over the Columbia river, Washington, USA	First (cable-stayed) segmentally constructed bridge in USA
Brecon Bridge, South Wales	
1980–1990	
Ely River Viaduct, South Wales	
Eastmoors Viaduct, Cardiff	
Stanstead Abbots By-Pass	
Grangetown Viaduct, South Glamorgan	Total length of 1040 m
Jeddah to Mecca Express-way, Saudi Arabia	More than 7000 segments bonded over a length of 20 km. Over 600 tonnes of adhesive used
River Torridge Bridge, Bideford, Devon	
Cogan Viaduct, South Glamorgan	Contains one span (of 6) with a length of 95 m
Post 1990	
River Lagan Bridge, Belfast	
Second Severn Crossing	

More commonly the joints are very thin, typically from less than 1 mm to 2–3 mm. This dimension, which depends on the close tolerance that can be achieved by matched counter-casting, obviously reduces resin consumption.

Where ambient temperatures have varied widely during construction, eg the Montreal Stadium and the Rio-Niteroi Bridge, both fast and slower curing resin formulations have been used (see Bibliography: Ciba-Geigy). In Montreal, cure was further assisted by infra-red heating, thus allowing work to continue down to -18°C; the resin temperature, and thus the progress of cure, was monitored by embedded thermo-couples.

Sharply contrasting colours for the two resin components are commonly specified in order to reduce mixing errors.

Pre-stress is usually applied across the joint in two stages: an immediate pre-stress (via steel bars) which is maintained for a short period, often 1–2 days while the resin attains a high degree of cure, followed by the permanent and full stressing of the tendons. Typical working stresses for segmental joints are generally considered to be limited to about 10 MN/m^2 in compression and 2–3 MN/m^2 in shear. These stresses are significantly less than the ultimate strength of the joint which, in general, is limited by the strength of the concrete and not the resin itself or the resin/concrete bond. Consequently, the allowable stresses for bonded segmental construction are generally similar to those for normal monolithic concrete.

Resin properties

Longer-term problems due to creep have not occurred (Hewlett and Shaw, 1977) even where a low modulus resin has been used, eg for Choisy-le-Roi a compressive modulus of 1.4–2.8 GN/m^2 has been quoted (Sims, 1979). Amongst other factors, the aspect ratio of the resin and the relatively low imposed stress assist in the limitation of creep.

When resin bonded segmental construction was introduced, relatively little information on the properties of resin systems was available. Consequently, extensive site and laboratory testing was carried out in association with many specific construction projects, see for example, Base (1962) for early work connected with the Sydney Opera House and also Hugenschmidt (1982). This showed that satisfactory rheology, cure rate under various ambient conditions, bond and short/long-term mechanical properties could be achieved. The continued use of this technique and the absence of problems on structures now in service for over 30 years have confirmed that the dependence on resin products was not misplaced.

The potential savings in time and costs have also been demonstrated by the projects listed in Table 3.5. In one specific case, Sydney Opera House, a six-month saving in construction time for the roof sub-structure was achieved (O'Brien, 1984) as the interval between joining and fully pre-stressing could be reduced by a factor of five when using resin in place of Portland cement mortar joints. More generally, cost savings of approximately 20 per cent have been claimed over alternative methods of construction (Mander, 1981).

In addition to the fast cure rate, epoxy resins also provide the following attractions for segmental construction:

- they lubricate the surfaces of the joints to facilitate positioning during erection (individual units for bridges may weigh up to at least 100 tonnes)
- cure shrinkage is negligible and the joints remain watertight, thus preventing leakage during duct grouting and, subsequently, corrosion of the pre-stressing tendons
- the thin joints are inconspicuous and aesthetically more pleasing than thick bands of concrete/cement mortar (and also essential to control costs)
- sufficient shear/tensile strength can be obtained for several units to be jointed and lifted into place as a monolithic assembly prior to pre-stressing.

Assessment

Due to the important contribution of resins to the technique of segmental construction, a proposed standard for acceptance tests and the verification of epoxy bonding agents was prepared by the Fédération Internationale de la Précontrainte and published in 1978 (FIP, 1978). This comprehensive standard deals with the following areas:

- the required physical properties (pot-life, open time, rheology, adhesion, cure rate and colour)
- the required chemical properties (chemical stability and durability in contact with concrete)
- the required mechanical properties (compressive, flexural and shear strength)
- the influence of temperature and humidity (and fire) on overall performance
- all aspects of site use
- a testing programme to assist the selection of appropriate products which covers 16 properties
- tests for checking the behaviour of the resin on site.

Further examples

Besides large structures, resin-bonded segmental construction has also been used to form slender columns, mullions, and roof trusses, for example at Coventry Cathedral and the terminal building at Glasgow airport (O'Brien, 1984). Several other forms of adhesively bonded joints which have some association with segmental construction are discussed in Section 3.8.

The future

As noted earlier, the main structural demand on the resin is the transmission and uniform distribution of compressive stresses; vertical shear stress is minimised by the use of pre-cast keys (which also assist correct location of the segments). It has been stated (Mays and Hutchinson, 1992) that the resin could act in a more structural mode and fully bear the vertical shear stress. However, reluctance arises from the 120-year design life requirement for highway bridges in the UK and the lack of sufficient data on the long-term durability of resin adhesives.

Such reluctance may be interpreted as "conservatism" as the semi-structural and waterproofing effectiveness of the resin joints has now been demonstrated over a period of 25–30 years. The very sudden collapse of the Ynys-y-Gwas Bridge in South Wales, after 32 years service, has certainly confirmed the potential ineffectiveness of alternative materials for segmental construction – tendon corrosion was attributed to chloride penetration through the sand/cement joints (Woodward and Williams, 1988). Alternatively, the reluctance may be seen as a wise precaution to maintain maximum safety factors and to eliminate extrapolation over long time periods.

Such opposing views are associated, almost inevitably, with many other demanding uses of resins in construction and have been particularly commented on by Hugenschmidt in relation to variations in formulation and the need for long-term testing (Hugenschmidt, 1982).

3.6 EXTERNALLY BONDED REINFORCEMENT

Reasons for use, scope and limitations

The need to strengthen (or stiffen) load-bearing elements in existing concrete structures can arise from one, or several, of the following factors:

- control of cracking where the existing reinforcement is inadequate due to design deficiencies or construction errors
- upgrading to conform with new or revised design codes
- changes in use which impose a load beyond the design limit
- damage of the existing reinforcement due to abnormal service, eg following impact (see below regarding corrosion)

These requirements are presented most frequently by buildings, eg columns, beams and suspended floor slabs, and bridge decks. They can be met by conventional improvement techniques such as partial/complete replacement, enlargement of cross-section (often with the provision of additional reinforcement), the application of external post-tensioning, or the construction of additional support.

An alternative solution is to adhesively bond plates to an external surface of the concrete element. Most commonly, one or more mild steel plates are bonded to the tension face to enhance flexural capacity (but occasionally to the shear or compression zones) using a specifically formulated two-pack cold-curing epoxy adhesive. More recently, glass or carbon fibre-reinforced polymer composite plates have come under increasing assessment as they have the attractions of high strength and stiffness, light weight and good resistance to fatigue, corrosion and chemical attack. They can also simplify surface preparation procedures.

The use of stainless steel is not generally recommended due to the particular difficulties associated with surface preparation (galvanised steel presents similar problems).

Externally bonded plates have been used for the repair of concrete structures affected by reinforcement corrosion. However, this can represent misuse of a technique which should be confined to strengthening, as in the circumstances itemised above. Where reinforcement corrosion does exist, repair techniques must be implemented to address this problem specifically and directly, eg the replacement of defective steel/concrete (see Section 3.10).

While plate bonding may then be used for further upgrading or strengthening, it should not be considered as a substitute for proper repair – use in this manner could lead to premature failure due to concrete disruption and de-bonding of the plate.

This technique has been investigated and used extensively throughout the world since its introduction in the early 1960s, as illustrated by an international seminar held at the University of Sheffield in 1990 (see Bibliography) which contained contributions from twelve countries. It is usually applied to reinforced (or pre-stressed) concrete but cast iron structures, which cannot be welded easily, have also been strengthened (Robbins, 1991). The technique has also been applied to timber beams, masonry walls subject to explosive blast and columns affected by alkali-silica reaction (Van Gemert, 1990).

Its viability and cost-effectiveness are well documented and, in many countries, the method is used routinely. Some guidance is provided in an FIP publication on good practice for the repair and strengthening of concrete structures (FIP, 1991). However, there has been some reluctance to approve and implement the technique in the UK., where reservations have been expressed by the Standing Committee on Structural Safety due to the lack of information on the long-term performance of both plate bonding and, also, resins more generally. In its tenth report (SCOSS, 1994) the Committee suggested that "a comprehensive review of case studies should continue" and notes that:

> "whilst some problems have been experienced due to corrosion of the steel, plate bonding has been widely used with no serious consequences known to the Committee".

As discussed elsewhere in this report, resins show a loss of strength as the temperature increases – particularly at and above about 50–60°C, the glass transition temperature for many ambiently cured systems. Consequently, in the UK at least, the use of bonded reinforcement has been restricted to highway bridges or other structures where the risk of fire is minimal and to buildings where, under fire conditions, the factor of safety provided by the existing structure is considered to be adequate. While this can limit the increase in imposed (or live) load capacity which can be attained, most requirements can be met in practice. Several authors (Hugenschmidt, 1975; Van Gemert, 1986) have suggested that insulated cladding could be used for additional protection against fire. However, tests to verify this point would be needed before such a procedure could be adopted.

The publication of a Department of Transport Advice Note (BA 30/94) covering reinforced and pre-stressed concrete members in highway structures should generate further confidence in the UK. Here, the need for bridge strengthening has been stimulated by an EC directive requiring a capability to withstand 40 tonne vehicle loads by the end of 1998. Increases in traffic volumes and vehicle weights (by as much as 40 per cent over the original design loads) together with a growing legacy of older deficient structures has created a similar urgent need to strengthen bridges in many other countries. For example, it has been estimated (Klaiber et al, 1987) that 40 per cent of the 600 000 bridges in North America require some form of rehabilitation or replacement.

Benefits

Compared to the alternative strengthening methods noted earlier, externally bonded reinforcement has the following advantages:

- it can be carried out with minimal interruption while the structure remains in use, eg bridges can remain open to traffic
- it is relatively easy to work around existing services
- temporary propping is not normally required
- there is only a slight loss of headroom
- the cost is acceptably low, the work can be carried out rapidly and maintenance is not excessively difficult.

Historical review and case histories

It would appear that the concept of externally bonded plate reinforcement was developed almost simultaneously in South Africa and France in about 1964–65. Two cases from South Africa have been reported: the strengthening of beams in the basement of an apartment building where the conventional reinforcement had been omitted inadvertently (Dussek, 1980); the emergency repair of a road overbridge following rupture of the reinforcement in several beams by accidental impact (Fleming and King, 1967). In France, a major bridge strengthening scheme on the Autoroute de Sud was carried out using plate bonding (L'Hermite and Bresson, 1967).

During the next ten years, or so, the technique was applied in Switzerland, Belgium, France, Australia and Japan and, in the following period, its use has become even more widespread. A number of detailed accounts of this growth are available (Hewlett and Shaw, 1977; Mays and Hutchinson, 1992; Mander, 1981; McKenna and Erki, 1994; Klaiber et al, 1987). However, it is worth noting that by 1975 over 200 highway bridges in Japan had been strengthened by bonded and bolted steel plates (Maeda et al, 1980) and that pre-stressed bridges were upgraded in the late 1970s in Belgium and Germany; by 1994, the figure for Japan had increased to 2500 (Department of Transport Advice Note BA 30/94).

It is also noteworthy that plate bonding has not been widely exploited in North America, where other techniques have been favoured for strengthening many bridges which have become deficient as traffic loads have increased (Klaiber et al, 1987). However, a number of examples connected with other structures, mostly utilising resin injection, have been discussed (Gaul, 1990). These include upgrading for seismic shock and, also, to maintain shear capacity following the formation of openings in walls.

The first application of plate bonding in the UK was carried out in 1966 when an eight-storey building constructed from pre-fabricated units in Essex was strengthened following a revision of the building regulations (Shaw, 1992; Hewlett and Shaw, 1977). This revision was stimulated by work at BRE on the effect of high wind forces on tall buildings (the collapse of the Ronan Point building following a gas explosion gave further cause for strengthening). Greater stability under very high winds was achieved by bonding plates to vertical columns and the central lift shaft.

Bridge strengthening/stiffening was first implemented in the UK in 1975 when four bridges, built in 1966 on the M5 at Quinton, were found to contain cracks due to a lack of tension reinforcement (Raithby, 1980). A total of 1376 plates (254 mm wide x 6.5 mm thick x up to 3.6 m long) were used.

As in other cases, the bridges remained in service and traffic was merely restricted while this work was undertaken. Tests using steel-to-steel lap shear specimens showed that the consequent minor strain cycling imposed during cure of the adhesive on site could reduce the shear strength by about 30 per cent. However, this was acceptable as the shear strength of the adhesive remained significantly higher than that of the concrete (Shaw, 1993).

Before and after load tests demonstrated the effective increase in flexural stiffness and the reduced crack opening under load (Raithby, 1980). It has been reported (Hutchinson, 1995) that after 19 years in service the plates at Quinton showed no significant signs of deterioration.

Subsequent UK applications in the 1970s included:

- M20/M25 Swanley Interchange Bridges in 1977 – plates bonded under and on the deck following flexural cracking due to inadequate deck slab reinforcement
- floor beams in a North London building in 1978 (increased floor loading due to change of use)
- pedestrian underpasses in Irvine New Town in 1978 (horizontal cracking in abutment walls).

Other examples performed in the UK between 1980 and 1987 have been cited (Mays and Hutchinson, 1988 and 1992). These include bridges, an underpass, various buildings and a multi-storey car park. The reasons for strengthening included:

- inadequate reinforcement – either for the design service conditions, increased loading resulting from a change in use or short duration abnormally high loading
- suspected fire damage to pre-stressed concrete beams (zero residual pre-stress was assumed)
- removal of roof beam support during reconstruction
- pre-stressed edge beams affected by accidental damage, eg impact.

Failure has been reported (see above) in only one case – the Irvine underpasses where adhesive failure after one year was attributed to a combination of inadequate surface preparation, lack of full adhesive contact and wet service conditions.

Shear strengthening is included amongst the examples given above. However, only flexural strengthening is permitted at present for bridges in the UK (but not abroad) as it is considered by the Department of Transport that further evaluation of shear strengthening is necessary.

3.6.1 Structural design

As the neutral axis of an element is lowered by external steel plates, hence reducing the fibre stress induced by applied loads, there is an optimum addition beyond which little, if any, strength increase is obtained. Unless the structure is jacked up prior to strengthening (to relieve dead load stress), the added reinforcement normally only carries imposed, ie live, loads. Thus, potential rather than existing over-stress is normally addressed. To prevent over-stressing of the existing reinforcement, working stresses in the added reinforcement must, therefore, be kept relatively low. In general, this is not an unacceptable restriction as steel plates are often used to ensure adequate serviceability by controlling crack opening rather than to increase the ultimate strength.

Due to this additional control of cracking, there were early concerns regarding sudden ultimate failure compared to non-plated elements, ie without the warning signs given by crack widening and possibly by horizontal shear failure in the concrete. Thus conditions which maintain ductile failure (plate yield) at the extreme should be provided as this enables failure loads to be more accurately predicted and allowed for in the design.

Prior to design, the existing condition and apparent problems must be properly diagnosed; thus detailed pre-design inspection (possibly with in-situ testing) is commonly required and, in some cases, mandatory.

Design calculations are usually based on normal reinforced concrete theory when steel plate bonding is used to enhance flexural capacity in an existing structure. The design procedure and detail recommended by Mays and Hutchinson (1992) are summarised in Table 3.6.

Table 3.6 *Steel plate bonding of reinforced concrete – structural design summary (after Mays and Hutchinson, 1992)*

Outline of design procedure

Calculation	Checking
(i) Stresses due to dead and permanent loads (existing section)	
(ii) Additional stresses due to imposed/live loads (strengthened section)	
	(iii) Acceptable total stress in the concrete* and each type of reinforcement
(iv) Interfacial shear stress in adhesive layer	
	(v) Acceptability**
	(vi) Adequate moment of resistance under ultimate conditions (with failure occurring in a ductile mode)

Design details

Recommendation	Comment
1. Plate thickness and aspect ratio	
Minimum of 4 mm+	Conformance with concrete surface irregularities requires reasonable flexibility but distortion during grit blasting must be avoided
Width/thickness of 60 for most satisfactory performance	Ensures yield failure at ultimate load and minimises risk of sudden springing
	Provides significant stiffening and increase in failure load
	Wide, thin plates reduce stresses but the edges of overly thin plates may lift due to out-of-plane transverse bending
Avoid ratios <40	Particularly in continuous beams at regions of hogging bending moment where shear/bond stresses in the concrete are likely to be relatively high
2. Plate length	
80–100% of the span	The complete length of the region in tension At least 80% of simply supported spans requiring flexural strenthening Elsewhere a minimum transmission length of 200 mm can be assumed
3. Bolting	
Designed to resist a shear stress of 3x mean value over the effective anchorage area (see also ref)	An adequate anchorage design should be based on adhesive bonding alone. Secondary anti-peel bolts are provided on an empirical basis
4. Cover plates	
Required to transfer load where plate ends are butted. Dimensions as the main plates. Total overlap length (for width/thickness = 60) of at least 400 mm	Joints must be minimised and avoided in areas of high deformation

Notes

* Although plating will increase the concrete area resisting flexural compression, the design may be governed by the need to limit stress within the existing section.

** Typical values of allowable mean shear stress for design under service conditions are 1.2 MN/m^2 for steel-to-concrete and 3.0 MN/m^2 for steel-to-steel.

Typical working shear stresses are 0.5– 1.0 MN/m^2.

\+ Increased from 3 mm since the publication referenced above.

The following points are also noted by these authors:

1. A typical epoxy adhesive correctly applied and cured under site conditions has an intrinsic shear bond strength of 10–20 N/mm^2. However, it is important that local shear stresses in the adhesive do not exceed the shear strength of the concrete anywhere (typically $\leq$4 N/mm^2). For steel-to-steel joints, formed by multi-layer or cover plates, a higher value of interfacial shear strength can be allowed as performance is not limited by a relatively weak substrate.
2. Bolts used for support during resin cure (and in the event of de-bonding) contribute to the plate anchorage although, with the geometry recommended in Table 3.6 their use is not strictly necessary. Their presence at plate ends provides a precaution against peel stresses but finite element analysis is required to fully assess their contribution.
3. More than one layer of plates may be needed to transmit the required load (while conforming to the aspect ratio recommendation for individual plates given in Table 3.6).
4. Many epoxies are likely to undergo a significant reduction in elastic properties and bond strength above 40–50°C or after long periods of immersion in water. However, mechanical properties are largely recoverable after isolated temperature excursions.

A recent brief review of an In-House design procedure (Farmer, 1994) also emphasises that high standards of workmanship and the use only of thoroughly tested materials are essential.

Design procedures are also covered in detail in the UK Department of Transport *Advice note for strengthening concrete highway structures* (BA30/94).

3.6.2 Requirements of the resin adhesive

A sample specification has been published which addresses the adhesives, bonding procedures and test methods for structural steel-to-concrete bonding (Mays and Hutchinson, 1988 and 1992; see also BA30/94). Requirements are specified to ensure acceptable structural composite action (ie the resistance of interfacial shear stresses) using a cold-curing epoxy adhesive. In addition to the main contents, summarised in Table 3.7, several points are made regarding the likely composition of the epoxy.

The adhesive specified above is defined as one which is capable of curing to the required strength of 10–30°C. A graph of cure time versus temperature is also required although a specific test method is not defined. However, a heat distortion test is included, primarily to determine the temperature resistance of the adhesive. It is suggested here that this test should also be used to assess the cure rate. This would avoid the anomalies which can arise when the more common strength measurements are used to assess cure at different temperatures (see Volume 3). The effect on cure rate of temperatures below 10$^{\circ C}$, as frequently met in the UK, could also be evaluated using measurements of the heat distortion temperature, thus indicating whether artificial heating is likely to be required on site.

Particular comment is made in this sample specification regarding the essential need for durable adhesion to both the steel and the alkaline concrete (and, therefore, resistance to the effects of moisture); also, for the particular need to carefully assess other factors such as: creep performance; differences in the elastic modulus and thermal characteristics of the steel, concrete and adhesive; health/safety requirements.

Table 3.7 *Steel plate bonding of reinforced concrete – resin specification * (Mays and Hutchinson, 1988 and 1992)*

Property	Recommendation	Test methods and comments
Form of material	Smooth paste-like consistency; initial viscosity of 20–150 Pa s at 20°C at a shear rate of 10^{-1}s; minimum yield stress of 20 Pa s at 10°C	Suitable for spreading on horizontal and vertical surfaces
	Resin and hardener components of dissimilar colour	To aid thorough mixing. Proprietary equipment is available to facilitate metering and mixing
	Maximum particle size (filler) of 0.1 mm	Filler should not settle out during the pot-life of the mixed adhesive
	Primers (corrosion inhibiting type for the steel)	Specified with the adhesive and suitable for concrete and grit-blasted steel
Working characteristics		
Application	Capable of being applied readily in 1–10 mm thick layers	Surface irregularities must be accommodated
Useable life	$\geq$ 40 mins at 20°C	Viscosity $\leq$ 150 Pa s within this time
Open time	$\geq$ 20 mins at up to 20°C	Time between adhesive application and joint closure. Difference between useable life and open time gives the time available for completion of mixing and spreading the adhesive. Exceeding the open time may lead to poor adhesion due to adhesive stiffening or atmospheric "contamination"
Storage life	$\geq$ 6 months at 5–25°C	Resin and hardener in original containers
Curing	Capable of giving the required strength when cured between 10°C and 30°C at 95% RH	See general comments in Volume 3
	The specified mechanical properties should be obtained in $\leq$ 3 days at 20°C	Negligible cure shrinkage should occur
Mechanical properties of hardened adhesive	Tests should utilise a storage/cure temperature of 20°C	Properties may undergo significant change if the adhesive is plasticised by absorbed water
Moisture resistance	Equilibrium moisture content $\leq$3% by weight Permeability $\leq$5 x 10^{-14} m^2s^{-1} at 20°C To enhance the maintenance of good adhesion even where moisture uptake does not degrade the adhesive	Immersion in distilled water at 20°C (1 mm thick specimens weighing at least 3 g are suggested)
Temperature resistance	Heat distortion temperature (HDT) $\geq$ 40°C. Primarily to ensure good performance under UK conditions and to enhance creep resistance	200 x 25 x 12 mm deep specimen under 4-point bending (maximum fibre stress of 1.81 MNm^{-2} as BS 2782). The HDT coincides with an additional 0.25 mm deflection using a surface heating rate of 0.5°C min^{-1}
Flexural modulus	$\geq$2000 MNm^{-2} at 20°C to assist in limiting creep under sustained load. $\leq$ 8000-10,000 MNm^{-2} to limit stress concentration effects.	Specimens as for HDT measurement. Loading rate of 1 mm min^{-1}
Bulk shear strength	$\geq$12 MNm^{-2} at 20°C. To ensure the adhesive is at least as strong as the substrate concrete and to minimise creep	Specimens and loading rate as for flexural modulus using a shear box
Tensile strength	$\geq$12 MNm^{-2} at 20°C	Dumb-bell specimens as BS 2782: Part 5: Methods 320 B/C, Cross-section of at least 10 mm x 3 mm. Strain measurements may be carried out
Mechanical properties of steel-steel joints	Adhesive thickness controlled between 0.5 and 1.0 mm	Detailed recommendations are given for the surface preparation of steel adherends and the fabrication of test specimens
Fracture toughness (tensile load acting normally to the crack surface)	$\geq$ 0.5 $MNm^{-3/2}$ at 20°C	Wedge cleavage specimens using a wedge insertion rate of 50 mm min^{-1} Similar specimens can be used for stressed durability testing

Table 3.7 (cont) *Steel plate bonding of reinforced concrete – resin specification * (Mays and Hutchinson, 1988 and 1992)*

Property	Recommendation	Test methods and comments
Lap shear strength	Testing performed over a range of temperatures, specifically including -25°C, +20°C and +45°C	Tests used only for comparative purposes in adhesive assessment, selection and control as failure load is influenced by geometry, rate of loading and test conditions:
	(i) $\geq$18MNm^{-2} at 20°C	(i) thick adherend shear test using a loading rate of 0.5 mm min^{-1} (recommended specimen differs from that in BS 5350: Part C5 to provide a more uniform stress distribution and to avoid peel stresses)
	or	
	(ii) $\geq$8 MNm^{-2} at 20°C.	(ii) double overlap joint which more closely reflects shear stress distribution in service
Fatigue	Constant frequency between 1 and 25 Hz with a sine wave form.	
	No failure after 10^6 cycles of a mean shear stress range between:	
	(i) 1.0 and 10 MN m^{-2} at 20°C	(i) thick adherend shear specimens as above
	or	
	(ii) 0.4 and 4.0 MN m^{-2} at 20°C	(ii) double overlap joints as above
Durability	Accelerated tests on various combinations of surface pre-treatments, primers and adhesives. Data collection from ageing in natural environments	For example, thick adherend shear specimens in a stressed condition and adhesive dumb-bell specimens subjected to appropriate conditions
Quality control	Manufacturer's normal test data for each adhesive batch Average results conforming with specification (no single result deviating by >15%)	Additional tests utilising site prepared/cured specimens are also recommended
Packaging	Correct proportions should be obtained by mixing the entire packed contents of each container	Recommendations are also given for labelling of the containers
Instruction sheets	Supplied with each delivery	Detailed contents are specified

It is noted that the environmental service conditions warrant careful consideration as the minimum required life for strengthening is 30 years. Consequently, the potential effects of a wide range of temperatures, condensing humidities and saline spray may need to be assessed.

The bond testing recommended in Table 3.7 is confined to steel-to-steel joints. The failure of steel-to-concrete test specimens is usually dominated by the relatively low cohesive strength of the concrete and, consequently, no useful information is conveyed on the ultimate performance at the adhesive/steel interface.

Where bonded reinforcement is used to upgrade steel structures, the strength of the joint is no longer limited by a weak substrate. However, a higher performance adhesive may be required as local ambient temperatures, particularly in box girders, can be high (Mays, 1985).

Strengthening by bonding steel plates, or other suitable materials, is included within the scope of products and systems for the protection and repair of concrete structures which is presently being addressed by the European Committee for Standardisation (CEN). This Committee and its associated working Groups (TC104/SC8/WG3 for structural bonding) are working towards the harmonisation of European technical standards, thus contributing to full implementation of the Construction Products Directive.

It is anticipated that some 20 or so testing standards relevant to plate bonding will gradually become available as drafts for public comment from approximately 1995 onwards. These methods deal with the working characteristics of the resin adhesive, its mechanical properties after cure, adhesion to steel and concrete, durability, fatigue under dynamic loading and creep in flexure.

3.6.3 Site procedures

For strengthening highway structures in the UK, site procedures (and a proving trial) are specified in detail in BA30/94.

The steel and concrete surfaces must be well prepared and free of all forms of contamination (chlorides, rust, grease, oil and dust) or weak material (laitance, mill-scale) in order to achieve and maintain good adhesion, the prime requirement of this technique. Cracks in the concrete which are wider than 0.2 mm and liable to leakage should be resin injected (see Section 3.4).

The mechanical treatment of concrete surfaces is normally advised and in-situ pull-off adhesion tests may be used to confirm adequate preparation (or concrete quality at the design stage). Steel plates are normally prepared in five stages, either on site or under factory conditions:

- initial abrasion to remove rust
- solvent degreasing
- grit-blasting
- dust removal
- priming.

Undue delay between the final stages has to be avoided to prevent rust formation (eg a maximum four-hour interval between blasting and priming is specified in BA 30/94). Both steel surfaces are often identically prepared to avoid errors in orientation when fixing the plates.

Although improved strength may be obtained from a direct resin/steel bond, a primer is commonly used as an anti-corrosion measure. The profile of the grit-blasted steel and the film thickness of the primer must be carefully specified for good bond strength. Additionally, the nature of the primer should allow bonding to be carried out effectively and with minimal preparation after lengthy storage or delays.

Any temporary support of the steel plates should be maintained until site-cured control specimens indicate that adequate resin strength has been achieved. Under some conditions, external heating may be required to assist curing. Finally, it is normally recommended that the coating on the exposed steel should be lapped over the exposed edges of the adhesive and onto the concrete.

The resin adhesive is generally applied to both surfaces. Immediate tightening of the support bolts then assists in the formation of an air-free adhesive layer, as excess resin is squeezed from the bond-line.

An alternative and economic technique utilising liquid epoxy resin injection of pre-positioned plates has also been used (Iino, 1981; see also BA30/94). This method can be of benefit with larger plate areas. Very fast curing resin systems can also be used, thus minimising the time of reduced traffic flow on bridges.

However, unless the prepared steel is properly primed (see above) corrosion may occur prior to completion of the injection. A foamed epoxy resin, placed by injection, has also been evaluated (Sawaide, 1981).

3.6.4 Research

In addition to numerous case studies, considerable research on plate bonding has been reported from many countries. Well-referenced reviews have been published by Klaiber et al (1987) and by McKenna and Erki (1994). The former has cross-referenced 26 papers published between 1967 and 1987 against 28 specific areas of investigation. The latter review deals in detail with research carried out at the University of Sheffield and at TRRL from the mid-1970s onwards. (Consequently individual references to this extensive work are not given here). McKenna and Erki also reviewed more recent work connected with the use of plates fabricated from fibre-reinforced polymers.

A significant part of this research has been concerned with the evaluation of resin adhesives and the development of appropriate tests (see, eg Table 3.7). However, extensive work has also involved large-scale flexural loading of rectangular and T-section concrete beams to determine the effect of the following parameters on deformation, ultimate strength, serviceability crack opening and failure modes:

- adhesive type, thickness and cure temperature
- imposed vibration during application and cure of the adhesive
- sustained loading and cyclic (fatigue) loading
- service temperature, thermal cycling and moisture effects
- plate dimensions, lapping and mechanical fastening
- the nature of existing reinforcement and concrete condition/quality
- long-term exposure to environmental conditions (and the effects of corrosion at the steel/adhesive interface).

Full-scale load tests on existing structures have been performed and some case histories have been updated by continued monitoring.

Although there is no unified and generally accepted code of practice for steel plate bonding, the above experimental work, associated theoretical analysis and continually increasing world-wide experience have all contributed to the confidence with which the technique is now employed.

A further boost in this confidence would undoubtedly be provided by standard accelerated test methods which could assist in the assessment of long-term performance and durability, particularly in externally exposed locations. Work in this area has been carried out, for example at the University of Sheffield. However, for the prediction of service life, the results of such indirect testing have to be related to externally exposed specimens or real structures – in the UK such work has been carried out at the University of Sheffield and at TRRL (see: Klaiber et al (1987); McKenna and Erki (1994)).

The use of plates fabricated from high strength fibre-reinforced polymers (FRP) was pioneered in Switzerland: on the Ibach Bridge, 6.2 kg of carbon FRP was used in place of 175 kg of steel plate (Meier, 1992). Glass FRP plates have been used in Germany on the Kattenbusch Bridge (McKenna and Erki, 1994).

FRP plates overcome two significant problems associated with mild steel:

- weight – due to handling difficulties the steel length is restricted and overlap joints are thus required; for an equivalent stiffness, a weight reduction of approximately 95 per cent can be achieved by using carbon FRP
- corrosion (particularly where de-icing salts are used) – this generates a need for very careful surface preparation/priming (and maintenance painting) and, also, uncertainties regarding bond durability.

Although FRP materials are expensive, their immediate and longer-term benefits are considered to give economic viability. They are also versatile as their physical properties can be tailored to design requirements (by varying the fibre/polymer composition and the fibre orientation) and their shape/dimension can be customised with relative ease. The incorporation of a peel-ply surface layer which is easily stripped off on site offers considerable potential for easing surface preparation (Hutchinson and Rahimi, 1993).

Research on FRP reinforcement (including the use of prestressed plates which can carry both live and dead loads) has been conducted in Switzerland, the USA, the UK and Germany. An extensive three-year collaborative programme, ROBUST is presently being carried out in the UK under the Government's LINK initiative (Peshkam and Leeming, 1994; Leeming and Peshkam, 1995; Leeming, 1996).

3.7 ANCHORING AND FIXING

Scope

The need to fix steel bars, dowels or bolts into a base material can be divided as follows:

- to anchor or retain an additional element, an added component or a new structure
- to stabilise the base material itself (or an existing structure) by imposing compressive stress via post-tensioning or by merely offering resistance to existing tensile/shear stresses.

The former includes such applications as:

- fixing of reinforcement starter bars which connect freshly added to hardened concrete (and anti-shear linkages between concrete or timber and mass or patch repair materials)
- fixing of an enormous variety of support or connecting brackets, plates, channels to horizontal, vertical and overhead surfaces – often of concrete, stone, brick or blockwork
- direct anchoring to the ground of large structures such as pylons, diaphragm and retaining walls, dock bases; in this case, steel cables may be used in place of bars
- anchoring of the tensioned steel used on cable-supported bridges.

The latter includes such applications as:

- stabilisation of rock formations against collapse (particularly during excavation) or to allow foundation or other loads to be accommodated
- stabilisation of old masonry and upgrading of deteriorated timber
- to assist in the retention of tiles, brick slips, facing walls or cladding panels
- tying of cavity walls.

In all these examples, reactive polymers may supplement or, more often, provide an alternative to the other common means of establishing a connection, namely mechanical action, casting-in-place or the use of cementitious grouts.

Resin-bonded fixings have been used widely for many years as they provide a number of specific benefits. This application is discussed in Section 3.7.1 and its extension to the in-situ replacement of corroded cavity-wall ties is discussed in Section 3.7.2. Two further applications are briefly discussed in Section 3.7.3: rock bolting for anchoring and stabilisation and cable (ground) anchors where resins are used both as grouts and, also, to fabricate the "dead end".

Although the specific end-use varies, the load transfer mechanism provides a common theme for these applications. Consequently, information (and references) given for one topic may be relevant to another.

The use of resins for compressive load bearing associated with anchoring, prestressing or support is dealt with in Section 3.9 and anchoring uses associated with repair are discussed in Section 3.10.

3.7.1 Bonded fixings

Applications and classification

Fixing assemblies (or "fastening systems") transfer concentrated loads into different base materials in a wide range of applications met in building construction, mining, foundation works, tunnelling, bridges, marine areas and on highways for holding down safety fences and parapets. These applications may include the formation of primary structural connections between prefabricated components and an existing structure.

Such fixings are closely associated in several respects with ground anchors and rock bolts (Section 3.7.3). A main distinction is the nature of the base material, ie concrete and masonry (most commonly) versus soil and rock.

Fixings may be classified as follows:

- cast-in-place: these are located in the formwork prior to placing concrete and are thus retained by passive mechanical locking, bonding or both
- post-installed: these may be retained in pre-cast or drilled holes by various forms of imposed mechanical action, eg expansion, wedging, screwing or under-cutting, or by adhesive bonding; some fasteners can be driven directly into the base material by an explosive charge.

Subsequent connections, made mechanically or by welding, may transfer load in tension/axial shear or in lateral shear (or both). In some cases, post-tensioning may be required.

The following discussion is focused on adhesively bonded fixings and their relative merits. However, a state-of-the-art report on the entire range of fastenings used with concrete and masonry structures has been published (Comite Euro-International Du Beton, 1994).

Standards for testing and technical approval

Although service rarely involves simple tensile loading, axial pull-out provides an effective means of testing fixings and is, therefore, most widely used. British Standard 5080 (Parts 1 and 2) for testing structural fixings in concrete and masonry covers tensile and shear loading.

More extensive and harmonised guidelines required for the European technical approval of anchors (metal anchors) for use in concrete are presently being drafted by EOTA. However, a UEAtc *Technical guide on anchors for use on cracked and non-cracked concrete* has been published by the European Union of Agrément as MOAT 49: 1992 (replacing MOAT 42, 1986).

In addition to design guidance, acceptance criteria and a statement of requirements, test methods for assessment are given. This represents a code of practice for testing prior to the issue of certificates of official approval (Agrément certification) confirming suitability for use with regard to safety, intended purpose and durability. In presenting test methods, this guide acknowledges that the influence of many factors is insufficiently known for load-bearing design to be based solely on a theoretical approach.

Types of bonded fixing

Bonding of post-installed fixings may be achieved with cementitious or other inorganic grouts, or with the following resin-based systems:

- glass capsules or "sausage-type" cartridges (see Volume 3) utilising polyesters, vinyl esters (epoxy-acrylates) or occasionally (meth)acrylates which are mixed in-situ
- partitioned cartridges fitted with special nozzles (see Volume 3) which induce mixing of the two components during pressure injection; both polyester and epoxies are available in this form of packaging (which is designed to deliver the correct mixing ratio)
- unfilled or filled pourable grouts, commonly epoxy or polyester, which are supplied as two separate components (or three where the aggregate/filler is not pre-mixed). Small volumes of two-component polyesters and epoxies can also be supplied in divided sachets which provide a pouch for mixing by hand.

Where post-tensioning is required, "end-anchorage" can be achieved by a combination of mechanical locking and adhesive bonding at the remote end of the fixing. Removable fixings combine internally threaded "block-ends" and isolation of the shank from the resin with a sleeve. Pre-stress is not normally applied to fixings which are bonded along the full embedment depth.

For bonded fixings in general, threaded or deformed steel is preferred; mechanical restraint then supplements the adhesion obtained on smooth surfaces. Under-cut/reamed holes can also provide mechanical locking.

The respective benefits of various post-installed fixing systems are summarised in Table 3.8.

Table 3.8 *The relative benefits of various post-installed fixing systems*

System	Advantages	Disadvantages
Mechanical	Immediate resistance to load. Removable. No fire risk during installation. Serviceable at high temperatures. Can be installed under wet conditions.	Considerable displacement can occur at relatively low applied loads. High expansive/point loads on base material during tightening and service. Performance may be poor in weak materials. Poor resistance to impact/vibration loads. Poor performance close to edges.
Bonded	Corrosion protection. No expansive/point loads on the base material. Improved stress distribution. Can be used in weak, voided materials. Good performance close to edges. Good reaction to impact/vibration loads. Predictable performance. Can be closely spaced. No "slip displacement" on application of service load. Removable if special rods/sleeves used (ie with internally threaded sockets).	Delay while system cures. May not bond well under damp/wet conditions. Careful preparation work often essential. Rough surfaces may be required for good adhesion. Performance at high temperatures may be poor. Possible creep. Limited shelf-life of some resins.
Cementitious	Low cost. No fire risk during installation. Active (chemical) corrosion protection. Performance at high temperatures. Use in highly voided base materials.	Slow cure. Mixing and placing operations.
Resins	Pre-packaged systems simplify mixing/placement operations. Passive (physical) but high corrosion protection. Higher strength and improved adhesion. Faster cure rate. Improved resistance to impact/ vibration loads. Improved stress distribution.	Effect of temperature on cure rate. More affected by high service temperatures (and fire). Health and safety requirements, including fire risk during installation. Higher cost.
Polyester	Fast cure rate (full load may often be applied within one hour). Use at low temperatures (-10° C poss). Availability in pre-packed forms.	Shorter shelf-life than epoxies. Fire risk during installation. Cure shrinkage limits size of annular gap to about 5–10 mm. Adhesion may be limited on damp or smooth surfaces.
Vinyl esters	As for polyester but possibly improved elevated temperature performance and adhesion in damp/wet conditions.	Similar to polyester but higher cost.
Epoxy	Low cure shrinkage – can be used in wide gaps. Improved adhesion. Specific formulations useable in damp/wet conditions (and underwater). Very low fire risk.	Slower cure, particularly at low temperatures. Use generally limited to >5–10°C. Not available for direct placement via cartridge or glass capsule. Relatively high cost.
Pre-placed cartridge or capsule	Convenience – mixing/placement operations simplified. Low wastage/economic benefit for repetitive small volume requirements. No cleaning operations.	Only available with polyester and vinyl ester. Matching of system to hole size and embedment depth required. Resin loss and possible poor bond in voided substrates. Relatively high unit cost.
Injection cartridge	Available for polyesters and epoxies. Convenience and versatility, eg for small but variable volumes. Pre-packaging of thixotropic products (for overhead use).	
2/3 pack systems	Larger variable volume applications. Fastening where significant voidage present. Deep embedment.	Site mixing.

Particular care is required where voids may be present within the base material, especially if pre-packaged capsule/cartridge systems are used. A number of methods can overcome the problem of relatively poor base materials:

- plastic mesh or "sieve sleeves" used with injected/pumped resin or cementitious grouts in voided, perforated, or hollow base materials. While also minimising grout loss, expansion of the sleeve provides restraint against pull-out
- the above sleeves or rigid perforated sleeves minimise grout loss in low density relatively weak, porous base materials, eg blockwork. The rigid sleeve may be used with a pre-packed capsule/cartridge.

Most types of fixings located in cracked concrete require special consideration. Two particular resin systems are available for this use:

- glass capsules which incorporate stainless steel fibres and which are used in conically or corrugated under-cut hole configurations
- an anchor rod which has expanding and bonded sections in an alternating sequence; torque is applied to activate the expansion prior to gelation of the resin.

Installation

Supplier's recommendations for the installation of bonded fixings must be fully observed. These may include the following:

- rotary percussive drilling so that mechanical keying contributes to the pull-out strength. Where holes have to be diamond drilled they should be roughened prior to bonding (or the safety factor should be increased – see eg Beveridge, 1976). Initial drilling of an under-sized hole followed by rotary-percussive enlargement is often recommended
- limitations for the annular gap size
- avoiding under-sized holes, eg due to worn drill-bits. Bond failure during rotary mixing of cartridge/capsule systems may occur following premature gelation induced by excessive frictional heat
- avoiding over-sized holes with cartridge/capsule systems – efficient mixing may not be achieved and insufficient resin may result in poor bonding
- careful cleaning/degreasing of the steel bar/stud
- complete removal of all detritus from the hole, preferably by vacuum, following loosening of compacted material with a brush and/or compressed air. Dust films can lead to poor adhesion; greater quantities mixed into the resin during installation (ie of capsules/cartridges) can cause excessive dryness and, therefore, a poor bond and low strength
- immediate washing of the hole following wet drilling (to avoid the "cementing" of any dust)
- limitations on the level of moisture present
- the use of special coupling systems which enable the bolt/stud (commonly chisel ended) to be driven effectively into pre-packaged systems by slow drilling, thus ensuring satisfactory mixing; the resin must then be allowed to cure prior to removal of the drive unit from the bolt/stud.

Note: poor mixing is a very common cause of failure.

- ensuring that mixing of pre-packaged systems (by drill) is stopped prior to any gelation of the resin
- controlled application of prestress to "end-bonded" fixings (fully bonded fixings normally require no more than a nominal torque for securing components firmly – excessive torque may be detrimental to the bond).

The reliable fixing of reinforcement starter bars is assisted by special glass capsule systems. The arrangement of the resin components ensures good mixing when smooth bars are installed with a hammering action (manual or power-assisted).

Where anchorage is critical, it is recommended that proof-testing is carried out.

Working loads and modes of failure

The ultimate pull-out load of fully resin bonded fixings may lead to one of the following results:

- cohesive failure of the base material
- cohesive failure of the resin
- tensile failure of the steel
- adhesive failure between the resin and either the steel or base material.

With proper installation and a suitable resin, ultimate failure should be confined to either the steel or the base material. As the embedment depth is increased, tensile failure of the steel generally predominates, although impractical depths may be required for this to occur in weak substrates. High tensile steel is likely to require a greater embedment depth than mild steel if failure of the base material is to be avoided at higher service loads.

Where steel failure does not occur, a mixed failure mode may be observed, ie substrate failure and resin cohesive/adhesive failure in a lower section of the fixing.

While axial pull-out loads can be calculated for base materials of known strength (assuming a conical failure plane) extensive design information derived from load tests is also available and should be used. In the case of pre-packaged capsule/cartridge systems, such information is usually provided by the supplier for a range of steel sizes (commonly M8 to at least M30).

Extensive work carried out by Paterson was published by CIRIA in the 1970s (Paterson, 1976). This dealt with short-term (axial and combined tension/shear), sustained and cyclic loading; some testing at elevated temperatures was also performed. Further work has since been carried out by many other workers.

For fully bonded fixings, applied axial loads are generally concentrated at the near end, rather than being uniformly distributed along the steel length (particularly as relatively rigid resin systems are generally used). This can lead to localised failure at or close to the resin/steel interface which can then progress (slowly or rapidly) along the embedment length. End anchorages utilise the strength of the base material more efficiently as the applied stress is distributed throughout a much deeper section.

It is important that displacement of the steel is recorded during load testing. Safety factors are influenced by the load at any initial failure ("first slip") but are usually based on the higher ultimate pull-out load – these loads are generally much closer for bonded fixings compared to mechanical fixings.

Bonded fixings are often used where cyclic or impact loads are imposed, eg holding-down bolts for machinery, warehouse racking. Good long-term performance can be obtained although safety factors may need to be appropriately increased over those used for static loading.

Creep

Creep of the resin may lead to loss of pre-stress or some loosening of fully bonded fixings carrying sustained service loads. Under normal ambient temperatures and recommended working stresses, creep strains generally stabilise fairly quickly. Consequently, re-tensioning/tightening is only likely to be required during the early stages of service.

More unacceptable levels of creep may be experienced if the imposed stress and/or the service temperature increase significantly; the possibility of premature failure due to creep rupture may also need to be considered. The magnitude of these effects will also be influenced by such factors as the specific resin type, the aggregate size and content, the width of the annular gap. Depending on the particular circumstances, acceptable creep performance may be maintained if safety factors are increased as the service temperature approaches the glass transition (or heat distortion) temperature of the resin.

Elevated temperatures

The influence of temperature on the short-term displacement and failure load of bonded fixings depends on the glass transition temperature of the resin and, therefore, has to be determined for specific systems. In some cases, performance will decrease significantly at temperatures of only 50°C; in others, failure loads may drop by only 10–20 per cent at circa 100–120°C (Comite Euro-International Du Beton, 1994). Periods at elevated temperature, especially prior to load application, may serve to increase performance if the resin post-cures to give a higher glass transition temperature than is obtained from more normal ambient curing.

It has been suggested that polyester bonded fixings under load should not be subjected for any length of time to temperatures above 80°C (Beveridge, 1976; Paterson, 1976). It is recommended here that specific advice is obtained for both short-term and sustained loading at service temperatures between normal ambient and 80°C.

The effect of relatively short excursions to a higher temperature will depend not only on the properties of the resin, but also on the embedment depth and thermal properties of the base material.

These factors will also influence the time to failure of loaded fixings under a sustained fire. Compared to resin-bonded fixings, some mechanical, cast-in or cement grouted fixings may support load for a longer period. However, failure may still occur as the substrate and/or steel properties deteriorate under the fire conditions.

Adequate performance can be sustained under fire conditions where the embedment is particularly deep if only the resin close to the surface is affected (Beveridge, 1976). However, the use of bonded fixings with an embedment of up to approximately 300 mm should be avoided in fire rated structures, eg to secure suspended ceilings, pipe brackets, sprinkler systems (Beveridge, 1976; Paterson, 1978).

3.7.2 The replacement of cavity wall ties

Although widely used (in the UK.) in the 1930s, and even earlier, cavity walls with metal ties became a standard form of construction after the Second World War. The metal ties covered by a British Standard since 1945 (BS 1243) are most commonly made of galvanised (or, less often, bitumen coated) mild steel.

It was reported in 1979 (de Vekey, 1979; Moore, 1981) that the life expectancy of these ties could be well under the notional 60 years expected for the walls (13–26 years for wire ties and 23–46 years for the thicker vertical twist ties). Consequently, cases of distress, formerly associated with specific conditions (poorly made ties, aggressive mortars or extreme exposure) were likely to become more widespread (Mathews, 1983; Simpson, 1990). Although collapse could occur (and has), particularly under high wind loads, the distress was most likely to be confined to loss of serviceability, eg bulging, cracking or leaking of the outer leaf (the most severe corrosion can often occur in the metal embedded in the outer leaf rather than that exposed in the cavity).

Four possible methods of reinstating an outer leaf have been presented by the Building Research Establishment (de Vekey, 1979; Moore, 1981):

- demolition and rebuilding using new corrosion-resistant ties
- one by one replacement of the corroded ties by removing single bricks from the outer and inner leaves
- filling the cavity with a polyurethane foam capable of bonding the two leaves
- the installation of (expanding) mechanical fixings or resin-grouted ties/stainless steel rods to re-tie the walls.

The replacement of ties by resin-grouted steel rods, which has obvious benefits compared with the other techniques, was described in further detail in BRE IP29/79 (de Vekey, 1979). This technique can also be used where design errors or poor work-manship have led to insufficient or inadequate ties, or where there are deficiencies in masonry veneers or brickwork cladding. The basic stages are shown in Figure 3.14 for work from the outer leaf – working from the inner leaf has access and making-good benefits but disadvantages due to disruption.

For the 75 mm embedment depth of 10 mm threaded studding (austenitic 18/8 stainless steel) recommended by BRE, a pull-out strength of at least 2500 N/tie should be obtained in brickwork. This gives a generous safety factor against the highest anticipated wind-loads on normal domestic structures (approximately x5 for the standard tie spacing).

Most forms of replacement tie for masonry, including those which are resin-bonded, are stronger than the traditional butterfly type. The consequent possibility of increasing the spacing width during replacement has been investigated (Tutt, 1988).

The initial work of BRE reported in IP 29/79 (de Vekey, 1979) utilised a standard polyester resin/glass capsule system for bonding the rod to the inner leaf. A later BRE Digest (257, 1982) which reviewed associated applications and alternative methods included: bonding with thixotropic resins placed with a sealant gun or via hollow ties; bonding with liquid resins used with specially designed hollow ties which incorporate special seals, as shown in Figure 3.15.

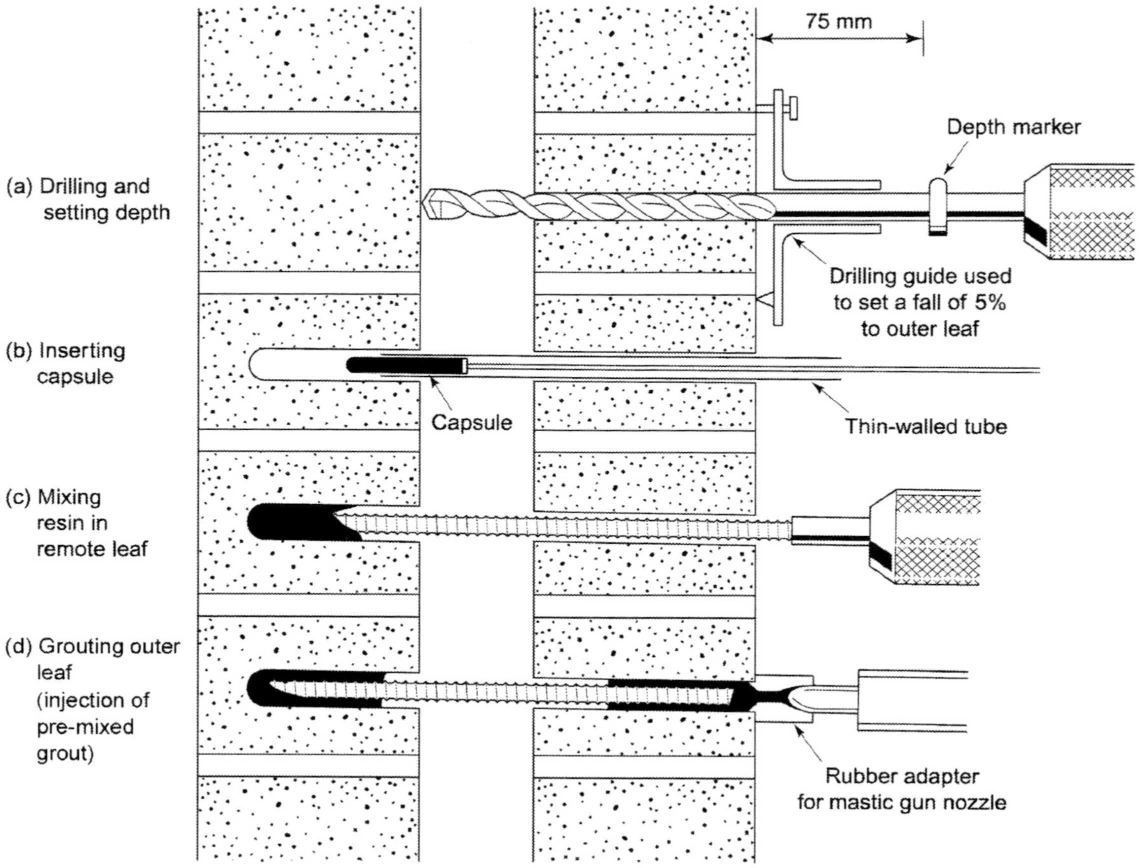

Figure 3.14 *Sequence of installation of resin anchors as wall-ties (Tutt, 1988)*

Since this early work, a variety of proprietary replacement tie systems have become available; some of these utilise GRP bars while others can be used in lightweight or hollow blockwork. Specially designed replacement ties which are mechanically fixed are less preferred to resin bonding in poor quality brick or blockwork (Paterson, 1994).

Polyesters are used most often for bonding ties due to their relatively low cost and tolerance of site use (mixing, proportioning, fast cure rate even at low temperatures). However, epoxy resins may be used in some circumstances, eg in damp/wet conditions or to bond GRP ties. The use of both has benefited from their availability in specially designed dispensing cartridges fitted with mixing nozzles (see Volume 3).

The appraisal of the existing conditions and the selection and implementation of a repair strategy for tied cavity walls are dealt with comprehensively in BRE Digest 329 (1993).

3.7.3 Ground anchors and rock bolts

Scope and applications – ground anchors

A ground anchorage is defined in the BS Code of Practice 8081:1989 as an installation that is capable of transmitting an applied tensile load to a load-bearing stratum. Many designs of anchorage, based on steel tendons grouted into bore holes to give both a fixed and a free anchor length, are used in rock and soil. These may be passive anchorages or they may be (and commonly are) post-tensioned. Load transfer to the ground takes place via the fixed length; over the free anchor length, the tendons are decoupled from the bore hole grout by grease and sheathing.

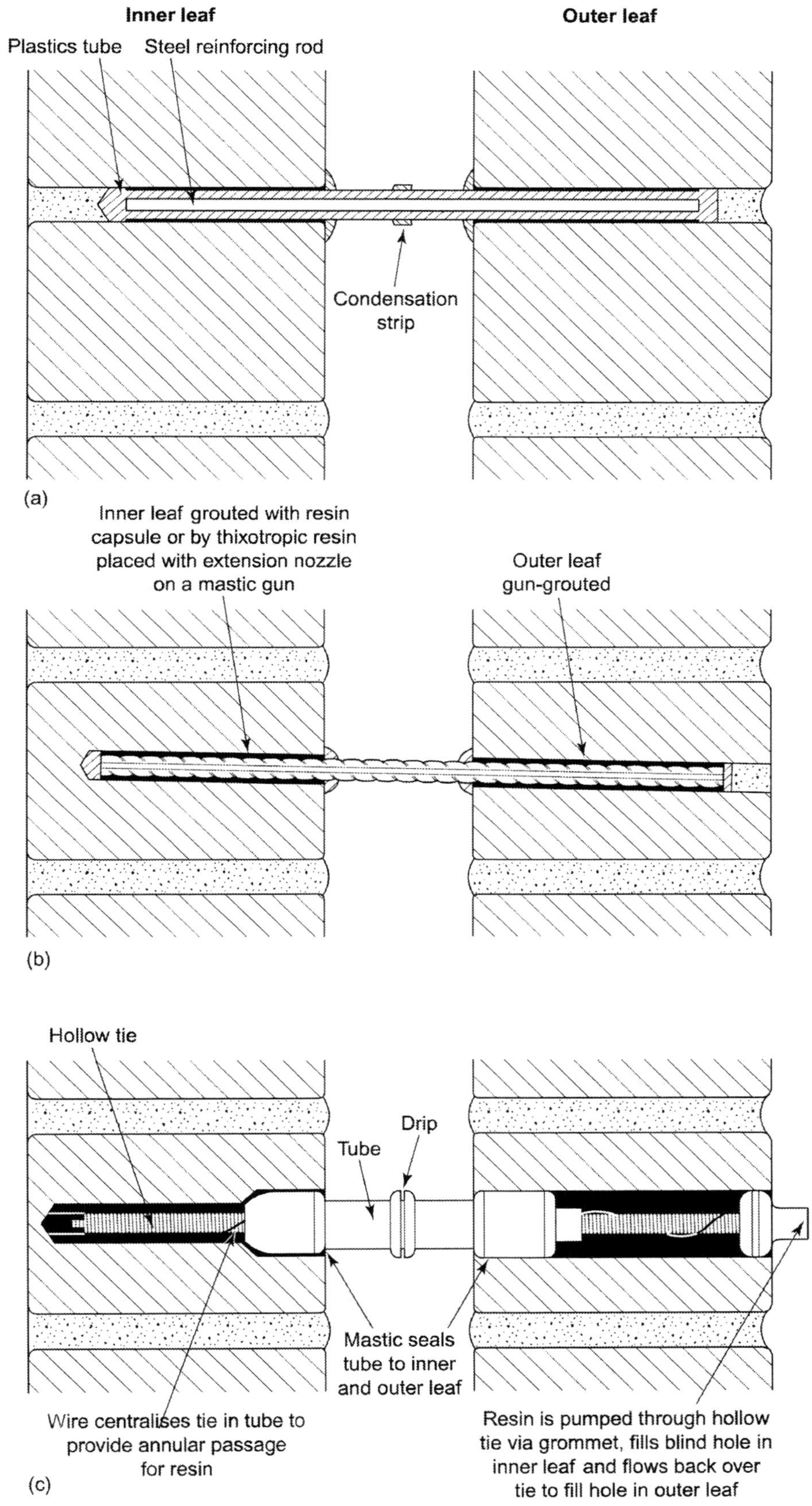

Figure 3.15 *Resin-bonded replacement wall ties (BRE, 1982)*

For temporary anchorages in soil, load transfer is usually affected via direct bonding between the exposed steel tendons and a cementitious grout. For permanent installations, the exposed steel tendons are encapsulated to provide greater corrosion protection. This encapsulation, which also serves to transmit load from the steel to bore hole grout, is fabricated prior to installation using either a cementitious or a resin-based grout. When permanent strand anchors are placed in rock, resin systems may also be used to grout the encapsulated (fixed) anchor length within the bore hole. Various forms of encapsulation are shown in Figure 3.16.

These anchorages are commonly associated with a very wide variety of structures such as retaining walls, dry docks, cofferdams, buildings, bridges and transmission towers.

Scope and application – rock bolts

Rock bolting, also dealt with in BS 8081, is a specific form of tensioned anchorage where a bar of steel (or GRP for example) is bonded into rock to increase its stability; for example, of existing/new slopes, or the surrounds to underground excavations – a technique developed in the mining industry which has been extended to allied civil engineering work. The former use is exemplified by the stabilisation of the rock faces below Edinburgh Castle in the 1970s (Gosney, 1974).

Although passive, untensioned anchorages ("rock dowels") are frequently used, the active type is usually preferred to effect a greater increase in shear resistance along discontinuities in the rock mass.

Resin systems, placed either as grouts or in a pre-packaged form (cartridge or glass capsule) are commonly used for post-tensioned rock bolting which may take one of two forms:

- the inner end of the bolt is bonded leaving a free length which is tensioned against a surface bearing plate, thus placing the rock under compression
- a fast-setting resin is used to form the fixed anchor zone, thus enabling tension to be applied prior to the setting of a less active resin which fills the remainder of the bore hole.

The second type, which combines permanent reinforcement with positive compression, can be formed in a single operation using a combination of fast and slow-setting capsules. As with fully bonded passive dowels, the resin provides good corrosion protection and performance is not impaired by spalling or other deterioration of the surface rock under the bearing plate (which can decrease the effectiveness of end-bonded anchorages).

Bonded bolts and dowels may also be used to stabilise old, deteriorated masonry walls or bridges and to anchor components or structures to a rock formation.

Resin types and properties

Polyesters, in "pre-packaged" forms or as site mixed grouts, are the most commonly used resins for anchoring and rock bolting applications, although epoxy or other types may satisfy special demands, eg a particular modulus requirement, adhesion to smooth or damp/wet surfaces, reduced post-gelation shrinkage. The main benefits of polyesters for these applications, including the fabrication of strand encapsulations on site, are similar to those discussed elsewhere in this report, ie relatively low cost and tolerance of site use and conditions (proportioning, mixing and rapid cure at low temperatures).

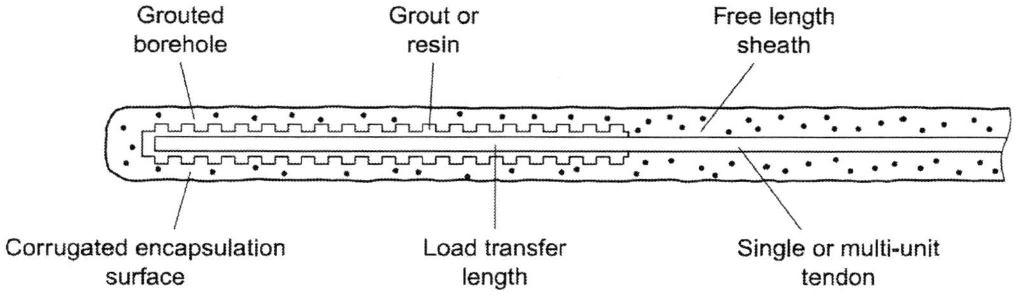

(a) Encapsulation with full length tendon load transfer

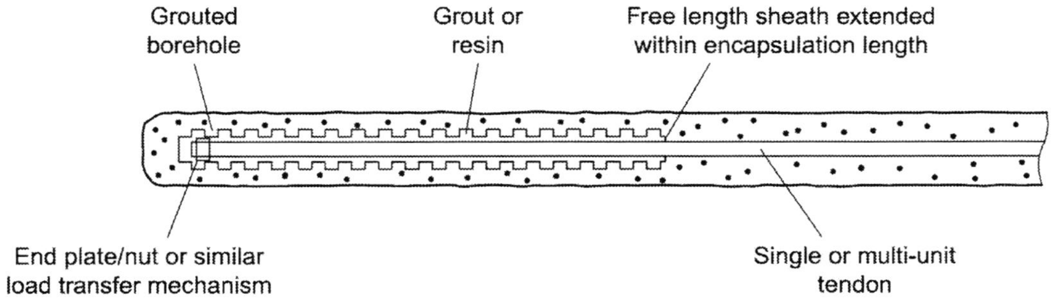

(b) Encapsulation with distal end load transfer mechanism

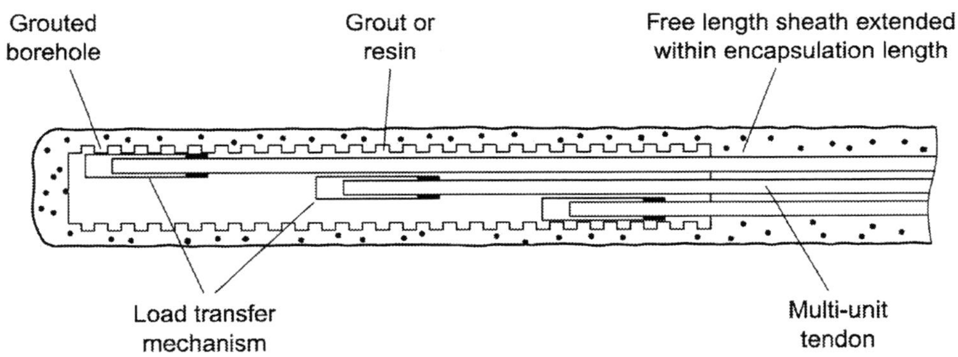

(c) Encapsulation with multi-unit load transfer mechanism

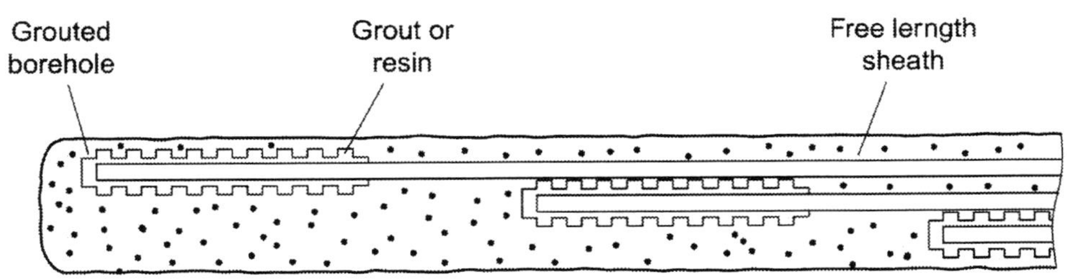

(d) Multiple encapsulation system

Figure 3.16 *Ground anchors – load transfer mechanisms for typical encapsulation systems (BS 8081, 1989)*

The basic requirements for the resin system are set out in BS 8081 where particular points are made concerning the relationship between strength/elasticity, post-gelation shrinkage, corrosion protection and creep.

While high strength is required for satisfactory load transfer, a low modulus and high extension at failure provide the following benefits:

- uniform load transfer along the bonded length
- compatibility with the load/extension characteristics of the steel, thus avoiding crack formation
- less tendency for pre-fabricated encapsulations to become cracked during handling.

The need to eliminate resin cracking in permanent strand anchor encapsulations and in grouted end-anchorage is essential to avoid leakage paths for water and the creation of potential hazards due to corrosion and/or loss of adhesion to the steel.

As the use of a low modulus resin may lead to unacceptable creep, there is a need to optimise these two properties.

Testing

Due to the relative absence of quantitative design data, particularly for resin encapsulated strand anchors, it is recommended in BS 8081 that, unless specific documented evidence is available, the evaluation of resin systems should include full scale proving tests. Ultimate load tests, creep monitoring and subsequent dissection of the anchor may be necessary to ensure suitable adhesive, mechanical and protective properties.

For the following reasons, care has to be taken to reproduce construction features when selecting or designing test procedures to assess the bonding element of anchoring and bolting: adhesion, frictional resistance and mechanical locking contribute to the pull-out resistance; the load transfer between resin and steel is generally non-uniform; the failure mechanism may vary and is determined by a number of factors; the ratio of free to fixed anchor length (together with detail at the anchoring head) can affect the imposed stress state, eg the surrounding material may be placed in varying degrees of compression.

Information

An extensive bibliography is given in BS 8081; some additional references are given in this report. In both cases, information on ground anchoring and rock bolting may have particular usefulness in the associated area of fixing (Section 3.7.1).

3.8 MISCELLANEOUS ADHESIVE APPLICATIONS

Background

The more prominent adhesive applications discussed in Sections 3.4–3.7 generally utilise resin systems (as unfilled liquids, pastes or mortars containing very fine fillers or aggregates) to form a relatively thin bond-line, ie less than 5–10 mm.

Related uses, outside the earlier categories, have continued to evolve and some typical applications are summarised here. The following Section (3.9) deals with applications where good adhesion remains essential, but where the resin thickness is often (although not essentially) greater than 5–10 mm, ie the resin would no longer be conventionally classified as a "glue".

Many of the earlier uses of epoxy resins have been discussed by Potter (1975). A more recent publication by Mays and Hutchinson (1992) deals comprehensively with all aspects of adhesive bonding in construction, including the potential for future developments. The range of applications summarised in Table 3.9 emphasises the versatility of reactive polymers conveyed in other parts of this report. It also illustrates their benefits in an industry which is unique, compared to other sectors which commonly employ adhesive bonding, as:

- adhesive joints usually have to be formed under unfavourable external site conditions rather than internal or controlled factory environments
- a wide range of substrate types and surface conditions are encountered
- common design lives of 50–100 years, in variable service conditions, generate a demand for very long-term performance data, particularly for primary structural connections
- the adhesive is often required to act as a gap-filler in addition to transferring stress. Existing design methods may be invalid for such joints as they were mostly developed for very much thinner bond-lines.

Table 3.9 *Adhesive applications in construction*

Application	Resin types most commonly used
Fresh concrete to hardened concrete (or steel)	Epoxy
Bonding tiles (steel, ceramic) and brick slips	Epoxy, polyester, furane
Glass-glass bonding (large stained glass windows)	Epoxy
Bonding of road-marker studs, kerb-stones and kerbing continuously extruded from fresh non-slump concrete	Epoxy
Pile splicing	Epoxy
Jointing of effluent and natural gas pipelines	Epoxy, furane
Steel sheet facing panels to pillars and walls	Epoxy
Bonding decorative aggregates, eg to panels	Epoxy
Joining of extruded aluminium window frames and aluminium balustrades	Epoxy
Bonding rubber strips over cracks or movement joints	Epoxy
The formation of thin compression or shear joints between sections of pre-cast concrete columns/mullions	Epoxy, polyester
Various joints for large prefabricated modular concrete units (Swiss system for single/multi-storey dwellings)	Epoxy
Bonded composite steel/concrete bridge decks	Epoxy
Structural steelwork connections	Epoxy
The connection of steel bar (reinforced concrete)	Epoxy

Resin types

Epoxies tend to predominant in adhesive applications, mainly due to their negligible cure shrinkage, wide availability in a variety of physical forms and capacity, when appropriately formulated, for bonding and curing well in wet conditions.

Polyesters are occasionally used for applications other than anchoring/fixing (Section 3.7) where only a compressive or otherwise relatively low stress is placed on the joint and advantage can be taken of their very rapid cure rate at low temperatures. The use of furanes is generally confined to similar types of connection where a particularly high degree of chemical resistance is essential, possibly at elevated temperatures.

While polyurethane based adhesives are common, their use in civil engineering is limited as most needs are met by the epoxy-based products readily available from mainstream suppliers in this field. However, polyurethane products from specialist adhesive formulators are used in a variety of applications which are associated with construction and building more generally, usually where a relatively thin "glue-line" is acceptable or required.

An extremely wide range of adhesives, based on both natural and synthetic polymers, is also used for bonding timber, decorative/protective finishes and other elements within buildings (see, for example, Mays and Hutchinson, 1992 and BS 5442, *Classification of adhesives for construction*). However, this area falls outside the scope of this report.

Several of the applications noted in Table 3.9 are briefly discussed below and the reader is also referred to the Bibliography.

3.8.1 Bonding new to old concrete

Fresh Portland cement concrete or mortar is frequently required to bond monolithically to mature concrete where structures or elements are patch repaired, rendered, or extended and, also, where thin or cement-rich toppings are placed on sub-floors. Even where stress has to be transmitted, adequate adhesion can be obtained provided the surface of the hardened concrete is thoroughly prepared to give a clean mechanical key (pre-wetting is often recommended, but no more than a saturated surface-dry state must be used as excessive water would lead to a weak interface (Bussell and Cather, 1995). However, such preparation is not always possible and is often very time consuming and difficult to carry out adequately under site conditions. Further uncertainties arise because of early shrinkage stresses, the relatively slow increase in bond strength and inadequate cement hydration at the interface due to drying out.

Provided the concrete has a good quality surface, free of laitance and contamination, appropriately formulated epoxy resins can give more reliable bonding (the fresh concrete or mortar must be placed while the resin is still tacky).

Water-based and low viscosity solvent-free systems can be used to provide an essentially adhesive connection; alternatively, a filled system of thicker consistency will also give some degree of mechanical keying of the fresh concrete or mortar. Where extensive shuttering has to be fixed, or where other delays are possible, resin bonding-aids with a very long open-time are available, eg up to 24 hours.

In recent years, aqueous dispersions of thermoplastic polymers and synthetic rubbers have been increasingly used to modify cementitious materials used for patch repairs, flooring, rendering or other surfacing work (see Appendix A1.5). With these proprietary or site-batched systems, a similar polymer dispersion is usually employed as a new-to-old bonding aid – typically in a cement or sand/cement slurry. Some slurries of this type, although not all, can dry very rapidly, particularly in hot, windy or low humidity conditions. They may then act as release layers rather than bonding aids.

This type of problem does not occur with epoxy resin bonding aids. However, these products should not be used in conjunction with polymer-modified cementitious systems (hardened or wet) unless specific advice has been obtained.

Epoxy bonding aids are particularly useful for relatively thin toppings on floors or other trafficked surfaces as an improvement in impact strength can be obtained.

3.8.2 Bonding glass, concrete, steel and composites

Glass

The formation of stained glass windows provides an impressive application of glass-glass bonding as seen, for example, in the lanterns of Blackburn Cathedral and Liverpool Roman Catholic Cathedral.

In one common technique, known as glass appliqué, pieces of stained glass are bonded to sheets of plate glass. Stained glass windows can also be formed from pieces of much thicker cast slab glass using a cement to replace the traditional lead cannes and water-proofing putty. Here, epoxies have often been preferred to cementitious materials as they are much lighter and have improved adhesion. For the Cathedral in Liverpool, about one ton of adhesive was used in this manner, the glass also being bonded within a concrete frame to give windows which could withstand 150 mph wind loads.

Concrete

An epoxy adhesive was used for pile splicing to form the 18 m long octagonal driven piles at the Melbourne Cultural Centre in Australia (Potter, 1975). A steel collar, sealed with a rubber-gasket, was placed around a protruding section and acted as a reservoir for the poured adhesive. After lowering the next pile section into the adhesive, a rapid cure (20 minutes) was effected with a steam-heated sleeve and pile-driving was then recommenced once the joint had cooled. This technique has been used in various parts of the world for both steel and pre-cast concrete piles. Its benefits for adjusting pile lengths to suit varying ground conditions have also been discussed (Heywood, 1974).

O'Brien (1973 and 1984) has quoted a number of specific examples, generally from the period 1960–70, where resin systems have been used for the jointing of pre-cast concrete structural elements. The maintenance of good performance in these cases was noted in 1984, although it was regretted that formal quantitative monitoring had not been possible, eg of creep. Five types of joint were listed.

1. Thin joints between post-tensioned units (further significant examples of this technique, segmental construction, are discussed in Section 3.5).
2. Thin compression joints between column and mullion sections.
3. Various bedding joints between columns, beams and slabs.
4. Compression joints containing adhesively bonded dowel connectors.
5. Shear connections in which bolts were used to maintain compression across the joint (beam/column connections).

Such applications may be classified as "semi-structural" because stress levels are relatively low, the resin is predominantly placed in compression and there are "fail-safe" arrangements. However, significant benefits are obtained with respect to appearance (ie very thin joints), speed of construction, effective and uniform stress distribution and water-tightness.

It was noted by O'Brien that very large safety factors compensated for the lack of knowledge concerning long-term properties of the adhesives. However, this information is essential if more structural uses are to be implemented. Short-term mechanical performance is not a reliable guide to long-term behaviour – both creep and creep rupture, for example, are dependent on service conditions such as temperature, the level and duration of applied stress and the formulation of the adhesive.

Steel-concrete

Both the actual and potential use of epoxy adhesives in the construction of steel-concrete composite bridges has been discussed in detail by Mays and Hutchinson (1992). Three methods are cited for achieving a bonded shear connection between the steel girders and the concrete deck slab, thus replacing the conventional welded mechanical fasteners.

1. Post-bonding where the pre-cast concrete deck panels are bonded directly to the flanges of the main steel girders to give a monolithic structure. This method has been used successfully in Germany, full-scale load tests showing acceptable strength and composite action after four years in service (see also Schutz, 1976).
2. Use of a grip layer formed by bonding coarse aggregate to a thin steel soffit plate; fresh concrete is then directly bonded to this rough surface. The steel plate also acts as external reinforcement for the concrete (see also Kriegh et al, 1967).
3. Pre-bonding where fresh concrete is poured directly onto an uncured adhesive which has been spread over the prepared surface of a flat or slightly curved thin steel plate.

The third method, also known as "pre-cast open sandwich panel construction", was developed by the Wolfson Bridge Research Unit at Dundee University (Mays and Hutchinson, 1992; Mays and Vardy, 1982). The composite deck slab, manufactured off-site, utilises the external steel soffit plate as a substitute for embedded reinforcement. Various methods may then be used to achieve continuity and composite action with the supporting steel girders, including adhesive bonding to the top flanges.

The combined steel/concrete thickness, of the order of 150 mm, gives a unit which is some 30 per cent lighter than a conventionally reinforced concrete slab of similar strength, a saving which increasingly reduces the cost of the supporting structure as the span exceeds 20 m (by 10–15 per cent for a typical medium span of 100–150 m).

The research associated with this concept included the assessment of a number of epoxy adhesives and also a polyester resin, the latter being rejected due to poor durability in damp conditions. Marked differences in performance were found between the epoxy formulations. In addition to short-term testing, extensive work was concerned with the structural durability of the bonded joint.

Trials under actual traffic conditions were initiated in 1980 on Dundee's northern by-pass and in 1985 a 6 m span bridge was constructed and placed in service. It has been suggested that this technique could also be used for floor applications or to replace existing bridge decks suffering from severe reinforcement corrosion.

Steel

For substrates such as concrete, glass, brick and ceramics, connection by other means, where possible, would depend on awkward mechanical methods. Consequently, adhesive bonding, usually with a high structural safety factor, is readily accepted. However, for structural steel-to-steel connections, bolting and/or welding methods can be used more routinely with predictable safety factors. Adhesive bonding, therefore, is relatively uncommon – in contrast to other industries where it has been used extensively for many years, eg the aircraft and automobile sectors.

In certain circumstances mechanical connections and welding present problems, eg:

- expense, practical difficulties or loss of design freedom
- distortion, stress concentration, joint slip, relatively low stiffness or fatigue failure.

Potentially, adhesive bonding could overcome many of these difficulties. One example quoted by Mays and Hutchinson (1992) utilises a special epoxy formulation, together with a novel surface pre-treatment, for the underwater repair and upgrading of offshore steel structures (as underwater welding is extremely expensive). In one other case, economy, speed and restricted working room led to the selection of adhesive bonding, rather than welding or riveting, for connecting existing patent steel floor arches to replacement edge beams on the Branbridges bridge over the river Medway in Kent. Here, bonding was achieved by injecting an epoxy between a steel trimmer plate and the flange of the new edge beam (Dumbleton, 1989).

Further cases of metal/metal bonding which have been examined, or which are under development, have been discussed by the above authors. These include:

- the fabrication of metal bridge parapets
- secondary steelwork connections on bridges, eg supports for concrete deck finishing machines, sign support brackets, lateral bracing connections, drainage pipe supports and girder web stiffeners
- the connection of stiffeners to steel plate structures
- various connections of structural steelwork (on bridges) where bonding with bolting can give superior fatigue performance to welded joints
- the fabrication of aluminium alloy beams and cruciform joints
- steel connectors or couplers for splicing rebars which combine mechanical action with adhesive bonding (see also Navaratnarajah, 1983).

Although further research and development is required, it has been predicted that adhesives could be used in the fabrication of steel frames for buildings before the end of the century (Oliver, 1994). The following benefits are foreseen:

- possible cost savings due to simplified, faster fabrication
- improved load transfer
- a reduction in problems arising from fatigue, brittle failure and corrosion.

Temporary structures and tubular space frames are considered to be forms of construction that may benefit initially from adhesive technology.

Composites

The adhesive bonding of fibre reinforced plates, for the external strengthening of reinforced concrete, is discussed earlier (Section 3.6.4) and the bonding of pre-cast resin concrete facing panels to lightweight core materials is noted in Section 3.12.

The 113 m long/63 m span cable-stayed footbridge at the Aberfeldy Golf Club in Tayside, Scotland, was innovative (in 1992) both for its all-composites construction and for its use of adhesive bonding. The 17.5 m high A-frame towers and the deck were assembled on site using a toughened epoxy structural adhesive to bond the mating surfaces of inter-locking GRP sections. The directly applied adhesive was dispersed from twin-cartridge packs using a pneumatic gun, thus ensuring rapid and thorough mixing and minimal wastage. It is anticipated that this bridge will have a service life of 30 years before first maintenance. (Ciba Geigy AG, 1994; Richmond, 1993).

In 1994, this technique was adapted to the construction of an 8 m span/4 m wide road bridge, the Bonds Mill Lift Bridge, which spans the Stroudwater Navigation Canal, near Gloucester (Head, 1994).

The future

The Standing Committee on Structural Safety of the Institution of Civil and Structural Engineers has highlighted the lack of information of the long-term performance of resins. It has warned against their use in critical locations unless effective replacement can be carried out should structural deterioration occur. The effect of the environment on resin adhesives over a very long design life, eg 120 years for bridges, is still unknown. Consequently, the adhesive bonding of steel may well be confined in the foreseeable future to secondary connections, to repair and strengthening (see, for example, Martin, 1992) and to use in conjunction with bolting. However, even without use for primary structural connections, more widespread applications may be anticipated as knowledge and expertise is further developed.

3.9 GROUTS AND MORTARS FOR GAP AND VOID FILLING

Applications which have an element of gap or void filling in addition to a varying degree of structural adhesion are discussed in Sections 3.4 (resin injection), 3.7 (anchoring) and 3.10 (repair). Additional uses outside the scope of these sections are dealt with here – usually the resin is used as a paste, putty, pourable grout or a trowellable mortar containing relatively fine fillers/aggregates. However, solvented epoxy "grouts" have been used to consolidate sand in the vicinity of oil-well bores (thus preventing blockage of the well pipes), to stabilise rock and to prevent groundwater ingress into mine-shafts (Potter, 1975).

3.9.1 Grouts for tiling and brickwork

Resin systems have been widely employed to grout ceramic or other industrial tiling used on floors and walls. Their resistance to chemical attack and water ingress is particularly useful for areas where acidic or other aggressive materials are handled. The retention of good adhesion and absence of cure shrinkage, given an appropriate formulation, also contributes to the maintenance of hygienic conditions (as the formation of fine gaps between the grout and the tiles is eliminated).

Such grouts are commonly based on epoxies although other resins have also been used including furanes, phenolics and special polyesters when particularly high chemical resistance is essential.

Resin grouts have also been used for external tiling, notably perhaps on the roof structure of Sydney Opera House (Potter, 1975). In order to exclude moisture and salts from the supporting concrete structure, some 130 km of tile joints on the chevron roof segments were grouted with an epoxy. This has been reported to be in generally good condition after more than 20 years service (an elastomeric joint sealant is used between the chevron units in order to accommodate thermally induced cyclic movement).

Resin grouts or mortars are also used for the jointing (or pointing) of acid-resistant brick linings. Sewer pipes within the extensive Cairo wastewater scheme, constructed from the mid-1980s onwards, provide an example of this application (Elliott and Spolton, 1985).

Here, a detailed and demanding performance specification for the mortar included the following properties:

- workability (assessed by site trials)
- coloured components that would assist confirmation of homogeneous mixing
- shelf-life, pot-life and rate of cure at different temperatures
- adhesion to brick, tile, concrete and cured resin mortar
- compressive/tensile strengths and flexural modulus
- water absorption and resistance to sulphuric acid and petrol
- cure shrinkage and thermal compatibility with concrete.

3.9.2 Pourable grouts for load-bearing applications

Concrete surfaces and grout properties

The precise location and support of heavy equipment on concrete surfaces can be very difficult, particularly where uneven cure shrinkage has occurred, as levelling tolerances of the order of 0.01–0.1 mm are often required. This operation can be simplified, with significant costs savings, by accurately positioning the equipment above the surface and then filling the gap (typically ≥about 10 mm) with a free-flowing (self-levelling) resin grout which bonds to both the concrete and the base.

Although proprietary cementitious grouts (usually expansive or shrinkage compensated) are increasingly available for some applications of this type, unmodified Portland cement grouts have a number of disadvantages, for example:

- relatively low strength and resistance to dynamic loads
- a relatively slow rate of cure accompanied by shrinkage
- poor resistance against acidic materials, oils and other chemicals used in typical industrial environments
- susceptibility to freeze-thaw damage.

Epoxy resin grouts are most common, mainly due to their very low cure shrinkage, although their excellent adhesion, relatively rapid strength gain, chemical/water resistance and good mechanical properties (high strength and good resistance to/ damping of continual vibration) are also important.

Both experience and testing have demonstrated that creep can be limited to an acceptable level with appropriate formulations. This is generally assisted by the low aspect ratio and (partial) confinement of the resin. Additionally, although high permanent (static) loads are often supported, the stress on the resin is typically only 10–20 per cent of its ultimate compressive strength (commonly 70–100 MN/m^2).

Particular consideration is required where relatively high service temperatures (above, say, 30–40°C) are likely for any appreciable period. The mechanical properties and creep resistance of some resin systems may then be inadequate, even at relatively low stress levels. Placing and curing at a similarly high temperature (or the subsequent controlled application of heat) may be beneficial if post-curing occurs (see Volume 3). However, this cannot be assumed and adverse effects may also occur, eg inadequate useable life, excessive exotherm, brittleness at lower service temperatures. In these cases, therefore, discussion with resin suppliers is essential.

Placement

Pumped or poured resin grouts have now been used successfully in diverse load-bearing applications for at least 30–40 years. Thus, the technique may usually be considered as routine with regard to the cured resin properties. However, specialist applicators should generally be employed to carry out the work as specific applications often have unique aspects. While merely "filling gaps" can appear to be straightforward, a poorly thought-out (or implemented) procedure can leave voids which are very difficult to remedy.

Good surface preparation is generally essential and, where the concrete is in poor condition, initial repair may be necessary. The underside of supported steel plates also has to be prepared to ensure good adhesion as tensile stresses will be imposed by vibrating or rolling loads. Heavy duty cranes, for example, create a "ripple" effect as they move along the supporting rails.

Reference has been made to mixing under vacuum, thus removing air which is entrained and displaced from the surface of fillers. If such air rises after placing, a very weak resin layer can be formed immediately below the base or sole-plate. However in well-designed formulations the entrained air will remain uniformly dispersed. Other than causing a small reduction in compressive strength, therefore, its presence is not detrimental. Where vacuum mixing is required, this should be clearly stated by the supplier.

Resin grouting requires attention to detail as its success depends on an appropriate combination of various factors, ie gap size, area, useable life and rheology; applied pressure and temperature (storage, ambient and substrates). With wide gaps and/or high temperatures, the cure exotherm can become excessive. In general, cost benefits are conferred by the fast cure rate of resin systems relative to cementitious grouts. However, at low temperatures, tenting/heating arrangements may be required. Site cured specimens may be used to determine when sufficient strength has developed for service loads to be applied.

Further examples

Several prominent and impressive examples of epoxy resin grouting are noted below (see also Bromley and Woodward, 1973; Hall, 1973).

1. Grouting of 25–40 mm gaps beneath the steel rails (50 cm wide sole plate) carrying a 60 m high crane at the Harland and Wolff Shipyard in N. Ireland. High dynamic stresses were imposed by the working crane which carried 800 ton steel sections along the 150 m wide x 750 m long dry-dock. Some 120 m^3 of grout was used.
2. Grouting of the gap between the foundation concrete and new sole-plates carrying the circular rail tracks for the Jodrell Bank Radio-Telescope. Accurate positioning and stability was essential following modifications which imposed overly high loads on the original concrete base. The telescope was fully operational within seven days of completing the work.
3. Grouting beneath the transporter crane rails at Parkeston Quay, Harwich (many other crane rail tracks at freight depots in the UK. have been resin grouted).
4. Grouting between the concrete wall and the back lining plates of the dock gates at Albert Dock, London and Seaforth Dock, Liverpool.
5. Grouting between steel-to-steel bearing surfaces on the Thames Barrier; this was considered to be the only feasible method of ensuring a uniform load transfer at a 20 m x 1.5 m interface between the barrier gate arms and the gates themselves (Tappin et al, 1984).

Railways

For many railway bridges in Europe and overseas, the conventional "trough and ballast" system has been replaced by a design which is similar to that of road bridges; a direct rail fixing method is then used on a flat deck. As rigid fixing of the rail base-plate, by bolting or welding, has several disadvantages, resin grouts have become widely used to form a flexible adhesive connection.

A tough resilient grout possessing some degree of compressibility is required and fast-curing two-component polyurethanes have been used since the 1970s. These grouts overcome problems due to unevenness in the base, absorb vibration and reduce noise emission. Stray current induced corrosion is also reduced by the high electrical resistance of the resin.

This method of fixing has also been used in place of the ballast and sleeper "floating" track system in metro and other rail tunnels. Here, in addition to the above benefits, the direct fixing method gives a significant saving in height and thus in construction costs (Reiner, 1990 and 1991).

3.9.3 Bedding mortars

Due to their rapid gain of high strength, resin mortars have been used for many years to form the support plinth, or seating, for bridge bearing pads. As the pads transfer load from the super-structure to the piers or abutments, the stress on the resin is mainly compressive. Epoxy mortars have generally been used but polyesters are acceptable in some circumstances.

The majority of the failures experienced during the 1970s were due to poor mixing of epoxy resin systems (Shaw, 1982). Generally, a compressive strength ≥ 65 MN/m^2 is specified; poor mixing can yield a resin which hardens but which has a compressive strength of <10 MN/m^2. An application guide for bedding mortars has been published by FeRFA (1991).

Polyester mortars and grouts have been used more extensively for the installation, raising and alignment of the frames for man-hole covers, particularly those associated with telecommunication lines – where at least 10 000 tonnes of resin was used in the UK for over 20 years from about 1970 (Raymond, 1990). A high compressive strength, eg 70 MN/m^2, can be obtained rapidly allowing installation to be completed in less than four hours and thereby minimising traffic disruption. The use of polyesters is usually preferred to other resins for work at low temperatures.

Resin mortars of various types have also been used for bedding machinery bases, runway lights, kerbstones and other street furniture, corrosion-resistant brick linings and tiles and brick slips.

3.10 REPAIR AND MAINTENANCE

3.10.1 Introduction

As traditional construction materials such as concrete, masonry and wood are inherently durable, they are routinely used in structures which have a very long design life. Given many different and often demanding service environments, some failures of these and many other construction materials in realising their full potential is perhaps to be expected.

That the incidence and scale of failures far exceeds levels which might reasonably be anticipated from the "ravages of time" may simply be attributed to "misuse", although the underlying reasons are more complex and include:

- the intrinsic complexity of the modern construction process
- insufficient attention being given to the design detail required for durability
- an inevitable tendency to generate unique combinations of performance requirements, material types, properties and service conditions
- the outdoor nature and scale of the activity
- misplaced emphasis on cost saving.

The general result of these and other factors is a national expenditure on repair and maintenance which closely approaches that for new construction.

In some areas, enormous repair budgets have been generated simply because knowledge and lessons from the past were forgotten or ignored. For example, the repair of reinforced concrete structures built mainly from the 1960s onwards now consumes approximately £500 million per annum in the UK alone, although the common causes of deterioration were known in the 1950s and even much earlier (Mays, 1992). This serious problem, recurrent throughout the world, has now, in effect, created a new and substantial branch of the industry.

For any incipient failure or loss of serviceability, there are, in general, three options (where some action is deemed to be necessary):

- demolition and rebuilding
- replacement of an element or component
- repair (which may include merely arresting the deterioration).

To minimise both costs and disruption, repair is usually favoured although, if cost-effectiveness is to be achieved, it is often the most technically demanding option.

Resin-based products are used in all manner of repair work due to the following attributes:

- they are available in a variety of physical forms and conform easily to most requirements in shape
- they bond well to many materials
- relatively rapid cure can be achieved under most working conditions
- they have good durability under a wide range of service conditions including very wet or chemically aggressive environments
- they are generally very resistant to penetration by liquids, vapours and gases.

However, they do have a number of disadvantages:

- a relatively high cost
- a need for skilled and experienced users
- a relatively low softening point (low glass transition or heat distortion temperature)
- some properties which can be significantly different to those of common substrates, eg modulus and coefficient of thermal expansion.

For these latter reasons, their use in some repair operations has tended to be replaced over the past 15–20 years or so by modified cementitious systems (see Section 3.10.2).

Common repair and maintenance needs for a number of traditional materials are summarised in Table 3.10, which also cross-references other relevant sections of this report. The main area not treated elsewhere, patch repair and surface void filling following the loss or removal of original material, is discussed below; in general, this discussion is concerned with hand-applied resin-based mortars and putties/pastes.

Table 3.10 *Reasons for repair and maintenance*

Defect	Resin-based repair: report section	Non resin-based material often used
Concrete		
Poor manufacture (mix design/placement/ cure) Excessive permeability	3.2, 3.4	✓
Cracking (non-structural)	3.4, 3.10	✓
Voids and blow holes	3.4, 3.8.1, 3.10	✓
Leaking construction joints	3.4	✓
Lack of cover	3.2, 3.3, 3.8.1	✓
Friable dusty surface	3.2	✓
Delamination/de-bonding of toppings or screeds	3.4	✗
Cracking (overload, lack of joints, inadequate design)	3.4, 3.6	✗
Cracking, spalling, delamination (reinforcement corrosion)	3.2, 3.3, 3.4, 3.8.1, 3.10	✓
Mechanical damage (impact, vibration, abrasion)	3.3, 3.4, 3.8.1, 3.9, 3.10, 3.13.2	✓
Chemical attack, freeze-thaw, salt-crystallisation and fire damage	3.2, 3.4, 3.8.1, 3.10	✓
Masonry		
Cracking	3.4	✗
Pointing	(3.9.1) (3.10)	✓
Need for consolidation	3.4	✓
Deterioration of ties and fixings	3.7	✓
Tiling and renders		
Inadequate joints (tile-tile)	3.9.1	✓
Cracking and loss of bond	3.4 (3.7.1)	(✓)
Permeability, chemical attack/weathering	3.2, 3.4	(✓)
Wood		
Splitting	3.4	(✓)
Rot	3.10	(✓)
Need for consolidation	3.2, 3.4	(✓)
Miscellaneous		
GRP (mechanical damage)	3.10	✗
Resin toppings (mechanical damage, chemical attack)	3.10	✗
Metals (corrosion)	3.2	✓
Asphalt (cracks, voids)	3.10	✓

3.10.2 The repair of concrete

Blow holes and surface levelling

Where relatively thin coatings are to be applied to concrete, blow holes (often more prevalent after surface preparation) and similar surface defects must be filled to eliminate sites of weakness and potential failure. Similarly, surfaces which are unduly rough (in relation to the film thickness) must be levelled – the material conventionally used is referred to as a "fairing coat". As the thickness of the final surface treatment increases from about 1 mm onwards, this preparation becomes unnecessary as the coating, lining or render itself is able to fill small voids.

Cementitious slurries, possibly containing fine sand and/or a modifying polymer dispersion (including proprietary products) are commonly used for surface filling and levelling. However, in some circumstance, a resin-based paste, putty or fine mortar may be more suitable, for example where particularly rapid setting/curing is essential. It must be ensured that subsequently applied materials will bond well to the filler/surface leveller. In general, any method of application suited to the consistency of the material may be used.

Cracks

Where "dead" cracks are present in concrete, and resin injection is not considered to be structurally essential, shallow sealing may be carried out. This may be achieved using low viscosity resins or water-based polymer dispersions which are merely brush applied or ponded on horizontal surfaces. These materials are suited to fine cracks; cracks wider than about 1–2 mm may be similarly sealed by "working-in" a cementitious grout.

Alternatively, but at greater expense, these cracks can be sealed more reliably by forming a chase with a width of at least 5–10 mm which is then filled with a well compacted cementitious or resin-based product; potentially weak feather edges should be avoided by giving the chase square or **slightly** under-cut sides. Where appearance is important, surfaces containing this type of repair have to be over-coated.

Where cracks are present in reinforced concrete, corrosion of the rebars needs to be considered when selecting a repair method.

Patch repairs

The above method for patching where cracking has occurred can be extended in principle to other damaged or deteriorated areas of concrete (see Table 3.10). However, as with any form of repair, pre-identification of the reason(s) for the problem is essential. It is particularly important to distinguish between damage caused by reinforcement corrosion and that due to other factors. This distinction (and the extent of the deterioration) is critical in determining whether patching is likely to be the most suitable and cost-effective repair technique. Several other differences also need to be considered, as noted below.

1. Whether the repaired area is structurally significant or is mainly cosmetic and thus carries a minimal proportion of the live load. In the former case the mechanical properties of the repair material and any mismatch with those of the concrete must be considered more closely, as discussed later in this section. For cosmetic repairs, it is desirable to achieve strain compatibility (or accommodation of differential strain) to avoid de-bonding or cohesive substrate failure; cure shrinkage, stress relaxation and differences in the coefficient of expansion are, therefore, important.

2. The depth of repair – as the depth increases normal resin-based (or cementitious) mortars (density 1500–2000 kg/m³) will tend to sag and detach from vertical and overhead surfaces. Consequently, the patch may have to be built up in stages with inter-layer priming. Alternatively (and up to about 75–100 mm) a proprietary light-weight mortar (density 1000–1500 kg/m³) may be used in a one-stage application. For still deeper repairs, Portland cement or resin concrete may be used (with appropriate formwork) to reduce cost and to avoid an excessive exotherm. In some cases, the concrete may be formed in situ by grouting/injecting pre-placed coarse aggregate (single size, 10/20 mm) using a cementitious or resin-based composition (Allen, Edwards and Shaw, 1993).
3. For large continuous areas, resin-based products with a higher filler/aggregate loading (and a different rheology) usually become more suitable than the pastes, putties and relatively resin-rich mortars that are commonly used for patching small voids. In addition to cementitious mortars and concretes, sprayed concrete is also likely to be considered for large areas.
4. Special techniques and materials are used for underwater patch repairs. It is essential that appropriate formulations (primers and mortars) are used for repair in any damp or wet conditions.
5. In one, or two, very specific cases, selection may be dominated more by the likely form of any failure than by the potential for durability. On jet air fields, for example, the detachment of sizeable pieces of resin mortar is much more hazardous than the gradual erosion of asphalt patching compounds.

The restoration of surface profile following relatively minor mechanical damage is usually straightforward, requiring very little pre-assessment other than a determination of whether any associated cracking needs to be addressed. Good workmanship is, of course, essential for a durable repair. The selection of materials for this type of patch repair is discussed below.

Where deterioration is due to fire, chemical attack, freeze-thaw effects or salt-crystallisation, patch repairs may be suitable but a greater level of initial investigation is generally required. Depending on the particular source of damage, this may include structural assessment, evaluation of the condition of the reinforcement and a determination of whether the residual concrete is contaminated or otherwise affected. Following this survey, the removal of further concrete (and the provision of structural support) may have to be specified as part of the patch repair process. A similar approach, ie cutting-out and patching, can also be appropriate for remedying poorly compacted, "honey-combed" concrete. In all these cases, the damage frequently extends over an appreciable area of floors, walls and external slabs. Thus renders, overlays and toppings are then required rather than patches.

Where cracking, spalling or delamination is (or may be) due to reinforcement corrosion, detailed preliminary investigations are essential for the development of a cost-effective repair strategy. The items which need to be addressed include:

- the reason for depassivation of the mild steel reinforcement, eg loss of concrete alkalinity (carbonation) or the presence of chloride ions (from manufacture or service conditions) or both; the extent of these effects also needs to be determined
- additional factors contributing to the onset of corrosion, eg insufficient cover or poor quality concrete due to inadequate mix design (by specification or use) compaction or curing
- the degree of corrosion (loss of steel section), its rate and its likely extent throughout the structure.

It cannot be assumed that patching of visible damage provides an acceptable solution where reinforcement is corroding as deterioration is likely to be progressive, particularly where chlorides are present. Hence, for cost-effective repair, the assessment noted above is essential, as is the consideration of other options which include:

- various combinations of patching, more major replacement of concrete (or whole elements) and surface treatment (for example, to reduce to the rate of carbonation); the replacement of defective reinforcement may also be necessary
- the use of special cathodic protection systems
- the application of more recently developed electrical/chemical techniques for the removal of chloride ions and the re-alkalisation of carbonated concrete.

Many aspects of concrete deterioration due to reinforcement corrosion are complex and require specialist knowledge for cost effective treatment. The brief discussion given above is a very simple overview; for more detailed authoritative guidance the Bibliography should be consulted.

One other type of repair may be viewed as an extension of patching and resin-based materials can be used: where there is access to the major part of defective concrete elements, for example columns or piles supporting a marine structure, repair may be carried out by complete encasement. Permanent formwork is initially positioned to leave an annular gap which is then filled with the repair material.

Materials

In certain circumstances, eg the small patch repair of non-reinforced concrete, high alumina and other special cements (magnesium phosphate, sulphur-based compositions) may be used. However, in these and other situations, resin-based patching mortars still provide a repair option, for example:

- for reinforced concrete where the cover provided by the patch is low (approximately 12 mm and less). Although cementitious mortars provide an alternative for the 6–12 mm range, a protective coating is usually advised to ensure adequate steel protection is maintained
- where a very rapid cure is essential, for example:
  - tidal areas
  - trafficked surfaces
  - where resin based coatings need to be applied very shortly after patching
  - military or other areas where return to service over-rides concerns regarding durability
- when it is difficult to properly cure cementitious systems
- where the temperature is very low, for example, cold stores (subject to a suitable resin formulation)
- to repair damage caused during condition surveys, eg from coring
- for unreinforced concrete, particularly where there is little variation in service temperature, eg floor slabs.

Due to their all-round properties, epoxy mortars are the most commonly used resin systems for patch repairs. The correct use of a compatible epoxy primer is usually essential to ensure good adhesion. Polyesters and (meth)acrylates are also frequently employed, particularly for cold conditions or where rapid cure is essential. The former generally need to be limited to small, shallow patches to avoid failures induced by cure shrinkage. Vinyl esters, furanes and polyurethanes are rarely used.

Patch repair usually requires a package of different product types, ie primers and/or bonding aids for the repair surface; coatings for the reinforcement; mortars for restoring the profile; fairing or levelling products and surface coating systems which often have a protective (barrier) and a cosmetic function. Generally, it is advisable to use one supplier for these products in order to avoid a split responsibility for their compatibility.

For durable, cost-effective repairs, (particularly of reinforced concrete) it is essential that the products in this system are compatible and that each component has the requisite properties. High compressive strength and a fast cure rate are insufficient and possibly misleading criteria for the selection of repair mortars. Modulus, the coefficient of thermal expansion, shrinkage, adhesion, electrical resistance and various barrier properties also have to be considered.

A European Standards Technical Subcommittee, CEN TC104 SC8, is currently working on the preparation of Harmonised Standards for the repair and protection of concrete structures. The work of SC8 includes performance-related standards for repair materials (including general principles for the use of products/systems), test methods and the agreement of minimum acceptable levels of performance for products/systems.

Portland cement concrete and mortar represent the simplest and perhaps most obvious type of material for concrete repair. While this is true in principle, the effectiveness of these materials in practice has been very varied; skill and attention to detail are essential at all stages of the work if the repair is to be durable.

Loss of adhesion, due to inadequate surface preparation and poor application and curing conditions (allied to cure shrinkage) has commonly occurred. Additional problems are caused by the mechanical weakness of thin sections, poor barrier properties, rapid carbonation and a relatively slow rate of cure.

Where these difficulties can be overcome, Portland cement mortars and concrete can (and do) provide a viable repair option, eg for concrete pavements. In some cases, adhesion may be assisted by resin-based or other bonding aids.

Resin-based mortars are attractive for patch repair work, despite their high cost, as they provide good adhesion, rapid cure and good resistance to weathering, chemical attack and mechanical damage. In appropriate situations, these products can give excellent service. However, they should not be used indiscriminately. Distress or failure can arise as a result of the cure exotherm and subsequent contraction, the relatively low modulus, the high coefficient of thermal expansion and the high electrical resistivity.

The latter property can be significant in relation to corrosion of the reinforcement and it is most undesirable when an impressed current cathodic protection system is also installed. In addition, while an "all-resin" patch provides good protection for the reinforcement, it is simply a barrier – the steel passivation induced by alkaline cementitious materials is not obtained.

Experience of these disadvantages, together with the high cost and essential need for very specific application skills, led to increasing use of cementitious materials from the early 1970s onwards, a transition that is now firmly established.

A key factor in this trend has been a greater ability to modify and tailor the properties of cementitious products using a wide variety of materials, eg polymer dispersions, micro-silica, cement replacements, fibres, corrosion inhibitors and chemical admixtures. Consequently, proprietary cementitious materials specifically formulated for concrete repair are commonly available, including fast-curing products with some form of compensation for cure shrinkage. Products designed for spray application or pouring/ pumping into a shuttered repair area containing exposed reinforcement have also been developed.

Structural repair – property matching

The repair of reinforced concrete often requires the supervision of a structural engineer as removing material to the depth of the reinforcement can lead to collapse of load-bearing members. The provision of appropriate initial support may, therefore, be necessary.

Concomitantly, replacing part of a structural member with a new material possessing very different mechanical properties will affect the manner in which service stresses are distributed. This will, of course, depend on where the repair material is used. Load sharing will depend on the relative stiffness of the repair material and the parent concrete. In some situations, Poisson's ratio effects can induce tensile stresses in the parent concrete.

For structural repairs, the following properties of the resin system must be considered (in relation to those of the concrete with which it interacts):

- elastic modulus – effect on load transfer
- compressive strength – to withstand load
- tensile strength/strain capacity – to resist cracking
- creep (stress relaxation), shrinkage and Poisson's ratio – in relation to load redistribution
- coefficient of thermal expansion – in relation to changes in load under varying temperature
- adhesion – to ensure load transfer.

Any incompatibility (between the respective properties of the concrete and the repair material) which leads to excessive differential strain will cause failure in the more brittle substrate.

Broadly, two experimental approaches have been used to investigate the effects of the mismatch in properties on the service performance of a concrete/repair material composite; however, each presents difficulties:

- individual intrinsic properties can be determined readily using standard methods. However, a complex model is then required for predicting performance in service – especially in the long term and particularly as geometrical and time dependent factors need to be taken into account
- various forms of composite can be subjected to simulated field conditions, eg temperature cycling. However, this type of work has, in general, given mixed results, often with poor repeatability or sensitivity; an "ideal" specimen geometry is difficult to define for relatively small scale tests.

It can be informative to combine both approaches, particularly when the results are supplemented by the following:

- field experience
- more limited evaluation of "full scale" specimens, prepared under various controlled conditions and then subjected to accelerated exposure regimes
- an understanding of the basic mechanisms governing performance.

More detailed discussion of this subject, including the results of experimental work, has been published by a number of authors (Kay et al, 1993; Emberson and Mays, 1987 and 1990; Mays and Hutchinson, 1992; Wood, 1990; Plum, 1990; Hewlett and Hurley 1985).

3.10.3 The repair of wood

While dry wood is an excellent structural engineering material, it becomes mediocre when wet, losing up to at least 50 per cent of its strength. Under certain moisture conditions, it is also particularly prone to fungal attack (dry and wet rot). Internally, this usually occurs when there has been some fault in construction or maintenance leading to dampness, a not uncommon occurrence, for example, where beam ends enter masonry walls. The destruction of wood by various boring insects is also common, as is the occurrence of shakes and splits.

When relatively modern structures suffer this form of deterioration, the timber is most likely to be simply replaced. However, as wood has been used in building for such a long period, there are many ancient and historic structures where restoration of decaying timber is preferred to replacement, in order to conserve the original form as far as is practicable. Resin-based products have an established role for wood repair here.

As with all repairs, it is essential to determine and eradicate the cause of deterioration before commencing any remedial work. With some old timbers, the need for repair may not be immediately obvious as an intact outer skin can conceal serious internal decay. Resin systems may be utilised in one, or a combination, of the following methods, in some cases to restore (or upgrade) the original load-bearing properties:

- replacement by a resin mortar (which may be reinforced with stainless steel starter bars)
- replacement of redundant sections with resin-bonded new timber (preferably of a similar age and quality)
- the insertion of resin-bonded reinforcing plates, steel or GRP rods or dowel-pins, eg to re-connect sound sections
- adhesive bonding of stainless steel plates (sound structural timber, eg flooring, may also be strengthened or structurally connected by this method)
- impregnation with a solvented resin, for example an epoxy or polyurethane (see Section 3.2)
- controlled low pressure injection of a low viscosity resin via drilled holes; epoxies, polyesters and (meth)acrylates has been used.

With all of these techniques, it is important that any preservative treatment does not affect the bond or cure of the resin.

A number of specific cases of timber repair have been given by Sharp (1981), Bowden and Shaw (1987) and by others (Roper, 1984; Bowden, 1984; Barrett, 1985; Anon, 1985).

More comprehensive accounts, dealing in detail with the design of repairs, resin formulation and a number of case histories, have also been published (Stumes, 1979; Phillips and Selwyn, 1978).

A UK research project supported by the Department of the Environment, "Long-term performance of resin bonded systems for structural timbers", was completed by TRADA Technology Ltd in 1995 (Mettem and Davis, 1996). In addition to case studies of in-service repairs, consideration was given to the classification and suitability of various techniques for load-bearing use and the material properties required for the design of repairs. The effects of moisture on the adhesion of epoxy systems was also investigated experimentally.

3.10.4 Miscellaneous repairs

Techniques utilising resin systems for the repair of brickwork have been discussed in earlier sections of this report: the resin injection of cracks, the consolidation of solid walls and the structural stabilisation of cavity walls with replacement ties (see Table 3.10).

General loosening of old brickwork constructed with lime mortar can also be strengthened by an unobtrusive "stitching" method. Here, a series of holes are drilled along the mortar joints of each brick course; resin injection at each location then bonds adjacent bricks, thus giving overall consolidation.

Resins have been used to bond the plates of restraint tie-bars to internal wall faces and to bond unconnected steel reinforcement plates to various masonry structures.

Resin mortars, based on epoxy resins containing carefully selected coloured sands and inorganic pigments have been used to patch repair deteriorated natural stone (Bowden and Shaw, 1987). Such systems usually need to be formulated to give a "sympathetic" repair in which the tone and texture of the natural stone is matched as closely as possible. Skilled application is often essential, particularly where intricate detail has to be reproduced.

Despite such care, the repair may still be visually apparent due to the different water absorption characteristics of the resin and the stone. With time, it may become more prominent due to different dirt-shedding and weathering properties.

Both epoxies and polyesters have been used as adhesives to directly bond replacement pieces to stonework, alignment and support often being assisted by grouted pins or dowels bridging the two parts (ibid).

3.11 FLOOR TOPPINGS

Areas of use and general benefits

Since their introduction to construction activities, thermosetting resins have been used extensively in various commercial and industrial flooring applications (see Sections 3.2, 3.4, 3.8–3.10 and 3.13).

The present section is concerned solely with resin-based monolithic toppings which essentially provide a protective finish to a structurally sound slab or deck/sub-floor (or screed) of concrete, steel or timber. They are used predominantly and routinely in demanding industrial areas and, also, for some commercial/institutional flooring, for example:

- chemical plants/factories and oil refineries
- nuclear power stations
- dairies and areas connected with food processing, brewing and the production of other beverages
- warehousing and other storage areas, including cold stores

- electronics and other manufacturing areas including engineering and plating workshops
- aircraft and automobile maintenance areas
- hospitals, laboratories, kitchens
- showrooms, foyers, concourses.

In these areas, many general requirements of a floor finish become critically important and less easy to achieve with directly finished floors or more conventional surfacing materials, for example:

- excellent resistance to penetration, attack or surface staining by a wide-range of aggressive chemicals, including hot materials, oils, hydraulic fluids and cleaning/decontamination solutions (assisted by an absence of construction/ contraction joints)
- very high resistance to wear and dusting due to abrasion and impact (often under heavily trafficked conditions)
- maintenance of good slip resistance (dry and wet surfaces)
- maintenance of rigorously hygienic conditions (assisted by non-porosity and relative ease of cleaning/decontamination)
- minimal maintenance and, therefore, long-term cost-effectiveness due to the absence of disruption during repair (following cracking, spalling, general wear, de-bonding from the base, loss of protective coating).

For many industrial areas, other floor finishes will give satisfactory service. For domestic and a high proportion of commercial/institutional flooring, selection will be governed by other requirements such as: appearance, sound insulation, low cost, comfort (some "give" or "feel" due to flexibility/lack of hardness) and relatively simple and rapid installation. In these cases, therefore, resin-based systems will rarely be used or considered.

Types of resin-based toppings

The resin-toppings considered here may be divided into two main categories.

1. Self-levelling (or, more correctly, self-smoothing) toppings, usually applied at a thickness of 1–5mm. These may be two or three-pack products depending on whether the aggregate/filler is pre-mixed or supplied separately and are usually classified as "medium-duty" systems.
2. Trowel applied systems with a higher filler loading which are suited to much heavier duty and are usually formed at a thickness of 4–10 mm. These are invariably three-pack products consisting of a resin, hardener or curing agent and filler.

Resin-based primers and, occasionally, sealers or coatings may also form part of the resin-based flooring system.

The self-levelling systems, which usually give a very glossy, decorative appearance, have to be manually spread to a near level finish; they then flow out to give the smooth topping. They are, therefore, somewhat similar to pourable resin grouts in consistency. For rapid application, or for very large areas, special formulations can be applied by two-component spray equipment fitted with heaters. More heavily filled flowing mortar systems are also becoming available. These give reduced laying costs and require less skill for a satisfactory finish in comparison with the more conventional trowelled products (Shaw, 1994).

Several sub-categories of these resin-based toppings are also available:

- toppings which can be steam-cleaned
- anti-static and electrically conducting products
- terrazzo flooring based on a resin binder
- self-levelling toppings with an element of thixotropy (thus providing a degree of "self-texturing" for slip-resistance)
- systems with a low odour during application – for installation in proximity to foodstuffs, beverages or pharmaceuticals.

Generally, a wide range of colours is available and special aggregates and fillers are also used for particular decorative effects (this includes coloured poly(vinyl acetate) flakes which are encapsulated in successive layers of a clear resin).

In the case of terrazzo flooring, the resin, most commonly an epoxy, simply replaces the Portland cement matrix of a conventional terrazzo finish. After trowelling and initial hardening, two water-lubricated grinding stages are separated by grouting to fill any pinholes, followed by polishing when a very smooth finish is required. The resin binder gives a higher performance floor at a much reduced thickness and a shorter period between installation and use.

Resin types

The following resin types are commonly used as binders for solvent-free flooring compositions:

- epoxies
- (meth)acrylates
- polyesters and some forms of vinyl esters, eg the styrene cross-linked urethane-acrylates (see Sections 2 above and Volume 3)
- polyurethanes (including one-pack moisture-curing systems).

As significant shrinkage accompanies the cure of conventional unsaturated polyesters, particularly careful formulation and laying is essential to avoid de-bonding/ delamination due to stresses at the substrate/resin interface (which can be exacerbated by differential expansion/contraction during service). Proprietary unsaturated polyester systems have been developed incorporating special additives whereby the shrinkage has been reduced or even eliminated, but, as yet, these have only found limited use (Rosenberg and Chapman, 1991).

A relatively high level of shrinkage also takes place during the cure of (meth)acrylate systems. However, the stresses are more readily accommodated as the resin remains in a thermoplastic state for longer, ie compared to polyesters, cross-linking occurs at a later stage of the hardening reaction.

There have been many attempts to combine the properties of reactive resin-based systems with those of hydraulic cement/water and a number of proprietary products have been developed with varying degrees of commercial success. The most successful has been a proprietary water-based polyurethane system containing Portland cement, which is available in a number of grades. These widely used products, both in the UK and overseas, can be applied to a concrete base after as little as seven days curing.

Water-based epoxy flooring formulations also containing Portland cement powder were described at an early stage (Makansi and Reader, 1975) and are still in commercial use. Novel polyester systems containing treated Portland cement and water have also found limited applications (Nutt, 1967, 1968 and 1969).

Selection and alternatives

Although industrial flooring is an established high volume application for resin systems, with well-proven benefits, selection of the most cost-effective products is often very difficult as:

- service conditions are frequently (and inherently) very demanding and varied with respect to chemical resistance and mechanical treatment; they are also difficult to quantify
- **useful quantitative performance standards, based on agreed test procedures, are presently lacking and formulations, therefore, are not produced to common and independent criteria; however, Harmonized European Standards are being considered by CEN**
- relatively few comparative assessments are published
- many products are available.

For many service demands, resin-based toppings are an obvious choice, eg for resistance to mineral acids, although experience and judgement is needed to select a product which satisfies the overall requirements.

In more marginal cases, where a cementitious topping, for example, could provide adequate initial performance, a resin system might be preferred in order to reduce maintenance. Here, the higher costs of materials are offset by future savings and, in particular, by the reduced disruption arising from repair work (see Section 4). It could be anticipated that this approach, placing a greater emphasis on whole-life costs, would have become more amenable to precise analysis as knowledge of long-term performance has increased. In fact, it has perhaps become more difficult in recent years as a bewildering number of options are now available.

Thus, for cementitious flooring, high performance and good durability can be achieved by the following routes:

- by good mix design (which may include admixtures), careful control of the water/cement ratio and close attention to the laying, finishing and curing techniques; monolithic granolithic-toppings may be included within this option
- ultra-high strength concrete which incorporates micro-silica
- by the use of surface sealers and coatings (see Section 3.2.2)
- specially formulated proprietary systems (see, for example, Harbron, 1994) which may incorporate a wide range of additives, eg polymers, plasticisers, micro-silica, special cements and shrinkage compensators
- by incorporating polymer dispersions, eg acrylics, styrene-butadiene (see Appendix A1.5) or a water-based epoxy resin.

While a detailed comparison of costs and properties would be particularly useful for flooring materials, this is beyond the present scope. Many options are available for a multitude of requirements and those particularly concerned with the selection of flooring should be assisted by a recent and detailed CIRIA report (Gatfield, 1997).

Depending on service conditions, and other factors such as cost and the time available for curing, the following additional alternatives to resin-based toppings may be considered:

- bitumen-emulsion modified cementitious floors, normally 12 mm thick, which are joint-less, dust-free and suitable for heavily wheeled traffic under dry conditions, eg in warehouses. Although subject to permanent deformation under high point loads, the underlying properties here may assist in avoiding the formation of cracks
- hot-applied mastic asphalt floors which, at 25 mm thickness, are very impervious with good chemical resistance (except to organic solvents) but are subject to deformation and corrugated wear
- industrial tiling (ceramics, quarries, steel) which can meet demanding requirements in terms of mechanical properties, speed of installation, ease of cleaning and chemical resistance (possibly assisted by the use of resin-based grouts and adhesives). However, the multitude of joints is a disadvantage in areas requiring a high degree of hygiene.

The assessment of service conditions

The selection or specification of a floor topping on a cost/performance basis can only be made satisfactorily if the service conditions and application requirements are clearly and precisely defined. Some compromise may have to be accepted where contradictory needs exist, eg ease of cleaning versus slip-resistance, low cost versus high performance. Some factors associated with the installation of resin toppings are discussed later in this section (see also Volume 2, CIRIA PR37).

The following service aspects also need to be carefully assessed:

- the exact nature, including concentration and temperature, of any materials that are likely to come into contact with the floor; merely specifying "good acid resistance", for example, is inadequate as the specific type of acid can be critical. The use of cleaning materials and accidental spillage must be considered – more routine spillage should be described in terms of the likely period of contact and the frequency/thoroughness of cleaning
- the falls and drainage that are available – in most instances resin toppings cannot economically alter existing levels
- the nature and concentration of traffic and any other severe mechanical action (including that imposed during plant maintenance)
- the service temperature range and its rate of fluctuation (including an indication of whether the flooring is extended externally)
- slip-resistance (specific legislation is relevant to some areas)
- the degree and ease of cleaning or decontamination required – compliance with EC directives and other statutory standards is essential in some areas, eg the food/beverage, nuclear and pharmaceutical industries, operating theatres
- the importance of maintained appearance. Due to their relatively high resin content, self-levelling toppings are usually very glossy and tend to emphasise scratch marks – unless initially dressed lightly with a slip-resistant grit or polished regularly (they also tend to mirror imperfections in the substrate which, therefore, requires a close tolerance surface regularity)
- the need to dissipate static electricity.

For some floors where extremely aggressive chemicals are occasionally spilled, a high performance but sacrificial finish can be considered, eg a high-build solvent-free resin coating on a resin topping. Resin sealer coats are also applied more routinely to some trowel-applied resin toppings for very high impermeability, thus assisting maintenance of strictly hygienic conditions.

Costs

Resin-based toppings (and sealers/coatings) invariably have a high cost. It is difficult to give precise comparisons due to the variety of products and other factors, eg degree of surface preparation required. However, 5–10 mm thick resin toppings will generally have a cost (laid) of £25–50/m^2, in comparison with approximately £10–15/m^2 for a 10 mm monolithic cementitious topping and at least £50 for high quality tiling (1994 costs). A comparison of the UK costs of different floor finishes, given in 1994 (Shaw, 1994) is summarised below.

Floor finish	**Cost (£/m^2)**
2-3 coats of inorganic surface hardener	1–1.8
2 coats of a non-reactive/reactive resin solution	1.2–2.5/2–3
1–2 coats of a chlorinated rubber	1.7–3.2
1–2 coats of a high-build epoxy coating	3.5–7
Self-smoothing epoxy (1–2 mm)	10–16
Polymer-modified cementitious (12 mm)	8–15
Epoxy/Polyester trowelled (6/6–9 mm)	18–24/15–20
Industrial tiles	15–30

The laying speed of different systems can also vary widely, but the self-levelling resin products compare well in this respect with any other high performance topping. Trowel-applied toppings, however, are significantly slower to apply (by a factor of about 5).

Curing and application characteristics

Specially formulated resin toppings, particularly those based on polyurethanes, (meth)-acrylates, polyesters, and vinyl esters (commonly, urethane-acrylates) cure rapidly, even at low ambient temperatures (0–5°C). Foot traffic is then acceptable after only several hours cure and full service after approximately 24 hours. However, many other products, the epoxy-based in particular, can require up to seven days cure at 15–20°C and considerably longer at <10°C to achieve full chemical resistance; some form of heating may, therefore, have to be considered.

The curing time of any system at a given temperature should always be checked with the supplier, particularly where early exposure to physically or chemically aggressive (or wet) service conditions is likely.

The application characteristics and cured properties of floor toppings can be varied widely by the formulator. The ease of use under common site-conditions, ie spreading, compacting and surface finishing is particularly important, as achieving a well-bonded and dense topping with a closed, flaw-free surface is critical for a high level of performance. Where requirements for ease of cleaning and decontamination are most extreme, the effect of formulation variables on surface micro-porosity may require detailed investigation.

Mechanical properties

Resin toppings generally have high compressive, flexural and tensile strengths (up to 100, 50 and 30 MN/m^2 respectively with certain formulations). Most importantly, toughness can also be very high, as indicated by a compressive modulus varying from 0.5–20 GN/m^2, an elongation at break of up to 15 per cent and an impact resistance considerably higher than that of concrete for example. Polyurethanes, (meth)acrylates and vinyl esters (eg urethane-acrylates) are usually outstanding in this respect. Most resin toppings will thus withstand direct and heavy impact without suffering the progressive distress shown by many concrete floors under physically aggressive conditions. Stresses at the resin/sub-floor interface, due to impact, will also be lowered as the thickness of the topping is increased.

Although a number of methods are available for the determination of abrasion resistance and many test machines for simulating wear have been devised, the quantitative extrapolation of laboratory results to service conditions is difficult. Such tests can, however, provide a ranking of different systems and a qualitative indication of the likely performance in service.

Developments in classification and testing (laboratory and in-situ) have assisted in the selection of concrete wearing surfaces, thus contributing to a reduced number of failures at one extreme or needless expense, due to over-specification, at the other (Chaplin, 1980; Kettle and Sadegzadeh, 1986; Concrete Society, 1994; BS 8204 Part 2, 1987). More limited work has demonstrated the significant superiority of self-smoothing and trowelled resin toppings compared to a 40 MN/m^2 concrete (Tillyer, 1990; Roberts and Smith, 1994). In general, well-formulated trowel-applied resin toppings can be expected to give a superior resistance to abrasion than most other finishes, with the exception of tiles and, possibly, ultra-high strength concrete.

A brief summary of unpublished laboratory work (Shaw, 1994) has illustrated the effect of the resin binder and, more critically, the filler type on the abrasion resistance of heavy-duty epoxy toppings. Hardwearing aggregates such as granite, calcined bauxite and cast iron grit gave particularly good results under the dry test conditions, as did a sand/gritstone blend. Particularly severe abrasion can occur under wet conditions, as exposure and slight wear of the fillers/aggregates can generate a very effective grinding paste which accelerates the loss of binder.

Chemical resistance

All resin-based toppings are conventionally described as "high performance" with regard to chemical resistance. This should not be taken to imply that excellent resistance will be obtained in all conditions and against all chemicals as good formulations in some applications may become poor in others. However, as formulations (and exposure details) are so varied, attempts to summarise potential performance can be more misleading than informative. Specific cases, therefore, should always be referred to suppliers, particularly where long contact times, high concentrations, hot materials or organic solvents/acids are involved.

Due to certain constituents, some common foodstuffs can be surprisingly aggressive, not only to concrete, but also towards some resin systems, eg the acids in fruit juices and dairy products, animal fats. Hence, it is essential to check that a product is satisfactory under these specific conditions.

As noted in Section 3.2, the assessment of chemical resistance can be misleading if the effect of aggressive liquids (including water) which diffuse to a coating/substrate interface are not considered, in addition to degradation of the coating itself. Due to their much greater thickness and the relatively high loading of fillers/aggregates, this caveat is less applicable to resin floor toppings – de-bonding resulting from spillage is usually of minor concern compared to direct attack (provided the resin system has been correctly formulated to give the essential impermeability). Chemical resistance can generally be assessed by determining changes in appearance, weight and a surface-sensitive mechanical property, eg flexural or tensile strength, following constant immersion of resin specimens (BS 6319 Part 8). An abrasion resistance test can also be useful as this property may be markedly affected by wet service conditions.

Specifying the most appropriate floor topping (concrete versus resin; one resin versus another) is often difficult when constant immersion tests show relatively slow chemical attack. Working practices then need to be considered, as service can be quite acceptable in these circumstances provided contact times are short and high standards of cleanliness are imposed.

For some special conditions involving contact or proximity to other media, specific requirements apply, for example:

- potable water: BS 6920 (1996)
- radioactive materials – ease of decontamination: BS 4247 Parts 1/2 (1981/82). Concrete, mainly due to its porosity, can be virtually impossible to decontaminate adequately, in contrast to resin toppings specifically formulated for this application
- foodstuffs, beverages: even without direct contact, some fully cured resin products may cause tainting of certain foodstuffs, particularly when ventilation is poor, as in cold stores for example (fats and dairy products are especially sensitive). Trials may be required to assess specific combinations (independent testing is conducted by the Leatherhead Food Research Association). Direct contact is restricted by legislation to a degree which varies from one country to another – as do the requirements affecting cleaning and hygiene (see, for example, EC Directive 93/43, Hygiene of foodstuff, HMSO). In this area, it is critical that specified procedures are followed in order to achieve full cure prior to normal service.

The effects of temperature and fire

The potential consequences of differential thermal expansion/contraction between resin materials and a concrete or steel substrate are discussed in Volume 3. Generally, problems are not experienced with resin-based floor toppings as:

- the coefficient of thermal expansion is reduced by a relatively high filler loading and particularly, therefore, for trowel-applied toppings (although coefficients are still significantly higher than that of concrete for example)
- many floor toppings are in an environment of reasonably constant temperature
- the thermal conductivity of resin systems is relatively low and, thus, small and fairly frequent ambient temperature fluctuations will not impose fatigue conditions at the resin/substrate interface (slower fluctuations may be assisted by stress relaxation effects).

Careful assessment is required for relatively high temperature fluctuations, particularly where the change occurs rapidly. Here the low thermal conductivity of the resin can be a disadvantage as the inherent differential in expansion/contraction is exacerbated by a temperature gradient, ie thermally, the substrate is relatively static.

Such conditions can arise in specific service environments or from elevated temperature cleaning operations. Steam-cleaning, for example, frequently leads to initial failure within the more brittle sub-floor followed by general breakdown.

Several approaches have been used by formulators to overcome this problem (Shaw, 1994) and, although some compromise in other areas may have to be accepted, resin toppings which can be steam-cleaned are available (a slightly greater thickness than normal may have to be used, eg 6–9 mm).

Where flooring is subjected to a constantly high temperature (eg close to floor-mounted ovens), the relatively low softening or heat distortion temperature of many common resin products needs to be considered, particularly where traffic is imposed or other stresses occur. Although some polyesters and polyurethanes may be useable up to (and beyond) 100°C, many epoxies are limited to 50–80°C.

Resin-based toppings can generally satisfy the fire resistance ratings (including spread of flame and smoke emission) which are imposed in European countries, although, in the UK, BS 476 does not deal specifically with such flooring. In some situations, particular formulations may have to be used.

Electrical properties

Low resistance, electrically conducting flooring is required where:

- the build-up and sudden discharge of static presents serious hazards (eg fire or explosion due to sparking where flammable materials are handled) or personal discomfort
- dust attraction is a particular nuisance
- sensitive electronic products or equipment can be affected.

Anti-static flooring often provides a convenient alternative to more conventional earthing arrangements as a means of reducing static charges to a safe level, particularly those generated by foot-traffic and mobile equipment (Richards, 1994; Tillyer, 1990). This type of flooring can include a range of products, such as self-levelling and trowel applied, slip-resistant, decorative, primers, sealers and coatings.

Special fillers/additives give sufficient conductivity for static charges to be dissipated harmlessly to earth, directly or via a minimum amount of earthing cable fixed between the charge source and the floor. Earth leakage grids may also be installed, eg between the primer and topping, to provide a high rate of discharge.

The surface resistance of flooring under specified test conditions determines whether it is classified as anti-static or conductive (BS 2050: 1978; see also BS EN 1081: 1998). In the former case, there is both an upper and lower limit for resistance (10^8 to 5×10^4 ohms) so that, while static is dissipated, the floor cannot become "live" via contact with faulty equipment or cables. Conductive floors have no lower limit and are suited to particularly and inherently hazardous areas where electrical equipment is routinely subjected to exceptional precautions. Thus, very effective static dissipation is of sole concern and the maximum resistance is equivalent to the lower limit for anti-static floors. Further, more general guidance on the control of undesirable static electricity is available (BS 5958 Parts 1/2: 1991).

A resin-based floor should ensure good long-term conductive performance due to the generally slow rate of wear and ease of cleaning. Due to its impermeability, it will also give (in contrast to concrete) stable conductivity, eg under variable humidity conditions.

Slip resistance

In general, resin-based flooring will give excellent slip resistance under dry conditions, while also providing ease of cleaning due to its smoothness and lack of porosity. However some degree of texturing is required for good slip resistance under wet conditions. This texture, which if too coarse will give undue cleaning difficulties, can be obtained from a scatter of fine aggregate on self-smoothing toppings (or on coated floors). For trowel-applied systems, carefully selected aggregate and fillers are incorporated at an appropriate level with coarser particles giving a slightly irregular surface to confer slip-resistance while impermeability is maintained.

The reliable in-situ measurement of slip resistance is very important as this property is initially somewhat dependent on application methods and can then change significantly with use (smoothing with wear or the build-up of contamination). It is also essential that acceptance criteria are agreed as slippery floors are the cause of many serious accidents.

There has been some controversy concerning the relative merits of two widely utilised devices for determining slip resistance, ie the Tortus and the TRRL pendulum tester (James, 1987; Bailey, 1988; see also Volume 3). However, BS 8204: Part 3 specifies the TRRL device for assessing polymer-modified cementitious wearing surfaces, requiring a temperature corrected slip resistance value (SRV) of not less than 40 in wet conditions (note: the SRV approximately equals 100 x the coefficient of dynamic friction). The correlation of pendulum SRV measurements with accident records for floors, stairs and pavings (Greater London Council, 1971) resulted in the following categories.

SRV	Condition
0–19	Dangerous – immediate attention required
20–39	Marginal – attention required
40–74	Satisfactory
75+	Excellent

Such categories may also be applied to resin-based toppings which, in due course, will be dealt with more fully in part of BS 8204.

Placement

Essential requirements for the use of resin products are discussed in Volume 2 and, more specifically for flooring, in the guidance document published in the UK by FeRFA, (1991). This guide contains extensive detail on very practical matters which must be observed if failures of a topping are to be avoided, for example:

- the existing floor must be structurally stable and sound (with a strength of at least 25 MN/m^2, preferably 35 MN/m^2 for heavy duty toppings) contain the necessary joints and have all appropriate falls and surface accuracy
- the quality of the substrate/resin interface is particularly important for the transfer of service loads without loss of adhesion and hollowness. Strong concrete floors can have a weak surface, eg due to over-trowelling. In general, a concrete substrate should give a tensile pull-off failure stress of at least 0.75 MN/m^2 after preparation
- good surface preparation is vital for good adhesion. All laitance and contamination, including existing coatings, and dust and preparation debris must be removed
- existing moisture levels in new or older concrete, rising damp and hydrostatic pressure require special consideration to avoid poor adhesion. Methods are available for determining the moisture content (FeRFA, 1991; BS 8204: Part 4).

For used floors, the existing condition must be assessed carefully **prior** to specifying preparatory treatment and materials. For new concrete floors, covering, rather than spray-applied membranes, should be specified to assist curing. The membranes will not degrade readily in internal conditions, due to the relative absence of weathering and UV light, and uniform and reliable removal is difficult.

Site conditions, including storage facilities, must be arranged for reasonable temperatures and the absence of contamination or interference from other trades during application and resin cure. The initial temperature of the resin and substrate must be considered, not just the ambient air temperature. The flow properties and surface finish of self-smoothing systems in particular can be affected adversely by low temperatures; trowel applied toppings can be difficult to compact thoroughly and to smooth to a closed surface texture.

Manufacturer's instructions for priming must always be followed. Trowel-applied toppings will not "wet" substrates very effectively and a primer is thus usually essential for good adhesion. Recommendations regarding the period between priming and application of the topping must then be observed. **With self-smoothing toppings, primers are required to prevent pinhole formation due to air-release from the substrate.**

Joints

Resin toppings are referred to as "seamless" or "joint-less" because, in principle, they can be applied in a continuous layer over large surface areas – in contrast to concrete, cracking will not occur during cure. However, it is essential that all movement and contraction or other joints in the sub-floor are carried through the resin topping. Specially cut chases in the floors and walls may also be required to avoid feather edges, eg at skirtings or where the topping has to finish level with an existing floor. As joints are particularly vulnerable to mechanical damage, penetration and chemical attack, careful detailing of arrises and selection/use of joint sealants is essential.

Osmotic blistering

In some relatively rare cases, involving finishes up to 6 mm thick, severe blistering has been observed between 3 and 24 months after laying. This has been ascribed to high osmotic pressures generated by water in concrete and water-soluble material at the interface when the primed/partially sealed concrete surface acts as a semi-permeable membrane. A more detailed discussion, including recommendations for prevention and rectification has been published (FeRFA, 1990).

3.12 PRE-CASTING

Areas of use

With only a few exceptions, for example factory-applied coatings, thermosetting resins are utilised in construction in three forms:

- systems which are mixed and cured on site, the main subject of this report
- pre-fabricated, fibre-reinforced composites (see Appendix.A1.1)
- heavily filled pre-cast polymer concretes.

On a worldwide scale, there is considerable resin usage in each of these areas. However, while on-site systems and composites are commonly and widely employed in most countries, the extent to which pre-cast polymer concrete has been adopted varies significantly from one country to another. Thus, it has been exploited extensively for many years in Japan, Germany and Russia for example, and more latterly in the USA; growth has been more recent in the UK.

The growth in use overseas can be followed through the published papers given at the ICPIC Conferences (1990 and 1992 in particular) and the various symposia organised by ACI Committees (see Bibliography). A monograph containing extensive detail on pre-cast polymer concrete, with a particular emphasis on Russian experience, is now available in an English translation (Mikhailov et al, 1992). Ohama (1978), Imamura et al (1978) and Okada and Ohama (1981) have outlined the early work in Japan where investigations commencing in the late 1950s led to the extensive use of polyesters, initially for cast man-entry chambers for communication cable lines. Later developments in Japan have also been reviewed by Ohama in the succeeding ICPIC conferences (see Bibliography). A comprehensive review of US and other specifications and standards has been given by Dikeou and Kaeding (1992); see also Ohama (1991).

Many of the applications of pre-cast polymer concrete reported in these sources and other technical and commercial literature are summarised in Table 3.11 where both structural and non-structural uses are included.

For panels and beams in particular, a number of variations are available:

- fully pre-cast elements which, in addition to mineral aggregates, may also contain fibre reinforcement. Various surface textures and patterns can be achieved. With appropriate design, such panels may be mechanically attached or resin-bonded to other components, in-situ concrete for example (note: temperature cycling needs to be allowed for in the design of bonded attachments)
- thin section pre-cast polymer concrete can be used as formwork which becomes a permanent facing for cast in-situ Portland cement concrete (with or without reinforcement)
- thin polymer concrete sections can be either bonded or cast directly onto a pre-formed core (including lightweight thermally insulating foam materials).

Performance

Although the properties of pre-cast Portland cement concrete can be varied significantly by appropriate mix design, pre-cast polymer concrete is, in general, more able to provide a superior performance in terms of:

- strength (compressive/tensile/flexural of, typically, 120/15/30 MN/m^2 respectively); thus, very thin sections which are not reinforced can be utilised
- strength:weight ratio (the relatively low weight of thin sections eases handling and installation)
- impact/abrasion resistance and good damping of dynamic loads
- resistance to chemical attack, water ingress and frost
- appearance (wide variations in colour and surface texture can be obtained directly, including exposed aggregate, with the expectation of minimal maintenance)
- rate of cure (benefiting mould utilisation)
- thermal insulation and electrical resistance
- dimensional tolerances and surface smoothness
- versatility in matching a wide range of specific requirements (by varying the type of resin binder or the aggregate)
- the fabrication of complex shapes (which can be cast with high precision).

Table 3.11 *Some applications of pre-cast polymer concrete*

Application	Comment and reasons for use	References
Large diameter (2–3.5 m) main sewer pipes	Furane resin-concrete used extensively in Prague (1970s) with overall savings in construction costs. Pre-cast elements provided an inner core for a cast in-situ Portland cement concrete outer shell. Excellent performance over >10 years shown	Bares, 1978
Highway gulley-pots	Extensive use in Holland since the early 1980s. Cast-iron gratings incorporated during casting (polyester). Benefits include: rapid production, reduced weight, erosion resistance, durability and improved flow characteristics	Raymond, 1990
Water mains (and sewer) pipes	Polyesters used in UK (70 km (>1000 tonnes) pa)	Raymond, 1990
Drainage channels	Testing and computer-aided design carried out in Germany	Zanke, 1989
	Interlocking polyester channels with integral cast-iron edge rails and gratings manufactured since the 1970s in the UK and continental Europe. Conformance to various Standards with high strength and excellent flow characteristics. Vinyl esters also used for chemical effluent. Falls can be "cast-in"	Raymond, 1990
	Gratings, traps, sumps and interceptors produced in the UK	
Pipes and heat exchanger tubes for geothermal energy plants	Special resin formulations post-cured at 180°C. Pre-cast pipes useable in contact with brine up to 260°C and 1 N/mm^2. Centrifugally cast lining for steel pipe useable up to 4 N/mm^2. Life cycle cost estimates showed potential benefits for a 20-year operating period	Kaeding, 1985 Fontana et al, 1987
Railway level crossings	Patented Bomac system successfully used in the UK. and continental Europe. Polymer concrete has replaced bauxite finished Portland cement (and granolithic) concrete with benefits, ie higher strength, reduced weight, electrical insulation, fabrication versatility, durability and maintained skid resistance	Taylor and Rayner 1989
Beams and panels	Structural investigation to assist optimisation of strength/durability/cost. Polyester concrete used to face sandwich elements with a core of lightweight Portland cement concrete, polystyrene foam, or rigid polyurethane foam. GRC also used as a core material	Fridley et al, 1989 Yaping et al, 1987
Light-weight panels for lining bunds around LNG storage tanks	Light-weight aggregates used to confer excellent thermal insulation properties and 6–40 MN/m^2 a compressive strength. Low water absorption beneficial	Fontana et al, 1987
	Also used as the facing for sandwich panels	
Building panels	Many specific examples available in USA, Holland, France, Germany, Belgium or Australia from 1958 onwards. Fibre reinforcement incorporated in thin panels (circa 30 mm) which are easily handled and installed. Simulated slate types available. Thermal insulation provided by a polymer foam core	Prusinski, 1978 and 1991
Blocks for floors and pavements	Marking requirements achieved permanently by use of differently coloured blocks	Pruskinski, 1978
Slabs	Used in Russia on hydraulic spillways to provide resistance to cavitation damage (approx. 2 m x 1 m x 23 mm)	Scanlon, 1981

Table 3.11 (cont) *Some applications of pre-cast polymer concrete*

Application	Comment and reasons for use	References
Miscellaneous	Street-lights, steps, plinths, cills, wall caping, window-frames, kerbing, decorative artefacts, waterproof ducting, amphibian road-tunnels, light-weight roof-tiles, lintels, electricity sub-station structures, ballistics panels, underground utility service boxes, railway sleepers, pipes with outer GRP skins, acoustic barriers, swimming-pool surrounds, platform edges, steps with cast-in electrical heating, radiant heating panels with cast-in hot water pipes	–
Corrosion resistant skirting	Vinyl ester, for example, used in dairies. Cleaning facilitated by smooth finish	–
Machine tool and other equipment bases	"Synthetic granite" widely used for bases/frames with improved noise and vibration damping compared to cast iron. Machine bases can be produced in several days, rather than months, due to fast cure and the ability to "cast-in" location points. Also used to support sensitive electronic equipment due to improved dimensional stability	ICPIC 1984, 1987, 1990 and 1991 (see Bibliography)
Simulated marble and onyx	Highly polished surfaces obtained via steel moulds and/or clear gel coats. 50 000 tonnes pa of polyester resin used in the USA in 1984 (with an increase in later years). Many other stone types can be simulated. Produced in Japan by at least 25 companies. Dense polyester and (meth)acrylate "marble" with a satin finish was launched by major companies from the late 1980s onwards. Widely used for domestic applications	–
Access chambers for communication, electricity and gas lines	Extensively used in Japan since 1967 (polyester). Approx. 20 000 tonnes pa used in the late 1970s. A total of 80 000 units installed between 1967 and 1984	Ohama, 1978 Imamura et al, 1978 and 1981
Highway traffic safety barriers	Polymer concrete shell filled on-site with OPC concrete	Laliberte, 1984
Formwork	Panels used as permanent shuttering for repair of bridge piers, central highway barriers and tunnels	Dikeou, 1987 Yamamoto, 1987

Compared to other more traditional materials, such as steel, porcelain/ceramics and engineering bricks, the relative benefits are more variable.

The properties of pre-cast polymer concrete which require more searching examination, particularly for load-bearing applications, are:

- modulus (values up to at least 35 GN/m^2 can be obtained)
- creep
- performance at elevated temperatures
- fire resistance.

Sufficient evidence is available to demonstrate that these properties do not exclude the structural use of polymer concrete on a general basis (see references given earlier). However, a detailed assessment may be necessary in specific circumstances, ie where the formulation and/or application is not already supported by testing and service experience.

Varied and often quite marked changes in properties are observed at temperatures above the glass/rubber transition (see Section 2 and, also, Volume 3). Within limits, short and isolated periods at a higher temperature may have little, if any, effect on serviceability; however, frequent or more prolonged exposure requires detailed assessment. At a certain temperature (generally above 150°C, depending on the particular resin and time of exposure) permanent chemical degradation will commence.

Flammability is usually reduced as the filler content increases. Self-extinguishing properties can also be improved by specific filler types, other additives or special grades of resin. The assessment of fire resistance should also take into account the particular form and thickness of the component. The emission of smoke and toxic products is also controlled by the resin type.

Structural use

Due to the increasing structural use of polymer concrete, the Society of Materials Science in Japan has issued a draft code, *Design recommendation for polyester resin concrete structures*, (Okada et al, 1987; Koyanagi et al, 1990). Although this code specifically mentions polyesters, the most widely assessed and used resin type in Japan, it can be applied to other reactive polymers which conform with its general requirements, namely:

- maximum thickness of 200 mm (which avoids problems caused by exothermic/ cure shrinkage)
- minimum compressive strength of 80 MN/m^2 at $\leq$50°C
- minimum modulus of 15 GN/m^2 at $\leq$50°C.

Individual assessment is required where there is variation from these values.

Polymer concrete reinforced by steel bars, pre-stressed tendons, glass fibre rods or fibre mats/cloths, for improved response to stress, is also covered by the code.

Linear elastic theory is adopted in the structural analysis, with serviceability and ultimate limit states being defined for four design classifications according to the imposed stress condition. Japanese Industrial Standards are available for the determination of properties required by the design calculations.

Work on the structural use of polymer concrete, including reinforced material, has also been carried out in the USA (Fowler, 1984; Yeon et al, 1987) and it is anticipated (Fowler 1991) to eventually lead to a further design code (see also Kaeding, 1991).

Some particular benefits of pre-casting

While pre-cast polymer concretes would be classified as high performance materials because they contain a resin binder, their claims are reinforced by a number of factors, some of which do not apply to site-cured resin products:

- pre-cast polymer concretes can often be cured at elevated temperatures, eg 50–100°C, with relative ease, thus allowing many properties to be enhanced, for example strength, thermal stability, weathering and chemical resistance
- high strength aggregates and specific additives (eg fibre reinforcement, fire retardants, shrinkage reducing agents and silane coupling agents for better resin/aggregate adhesion) are commonly used to improve performance

- formulation is not restricted to those resin binders which can be readily handled and cured under variable site conditions; additionally, good adhesion to different substrates is not required and exothermic reactions can be readily controlled. A greater range of options is, therefore, available to the pre-caster who, in general, also has more opportunity to formulate to very specific and clearly defined service conditions
- the pre-cast manufacturing process facilitates: very high aggregate loading (only 6–7 per cent binder by weight is used in some systems) and interlocking, or close contact, of aggregate particles; the use of extremely dry aggregates; good compaction with the elimination of voids; overall consistency.

Formulation

Polyesters, vinyl esters, (meth)acrylates and epoxies are the most commonly used resins for pre-cast polymer concrete; furanes and polyurethanes are used on occasion, as are some special hybrid formulations. Due to their overall performance and lower cost, polyesters are used predominantly although the use of (meth)acrylates is increasing; epoxies and vinyl esters tend to be used more for special needs.

High filler/aggregate loadings are assisted by the low viscosity of the (meth)acrylates and polyesters. The low cure shrinkage of the epoxies is of particular benefit when casting to extremely precise dimensions.

The use of polyesters, vinyl esters and, to some extent, (meth)acrylates lowers cost but the cure shrinkage then has to be allowed for in the mould design (high filler loadings and certain additives assist by reducing shrinkage). Additionally, these resins carry greater fire risks in the uncured state (processing equipment for (meth)acrylates usually has to be flame-proofed).

Vinyl esters and furanes are most commonly used for service demanding high chemical/temperature resistance. Furanes are favoured in some countries on economic grounds, eg South Africa, as large quantities of furfuryl alcohol monomer are produced from agricultural waste.

Where appearance of a high quality must be maintained, both resins and curing agent/catalyst-initiator systems must be carefully selected as some materials can lead to discoloration. Unfilled resin gel coats (tinted or clear) are often used to improve surface finish and weathering resistance – they can be applied initially to the surface of the mould. Ultra-violet screening agents may also be incorporated.

The selection of aggregate type, size and grading and the attainment of an optimum (usually very high) loading are most important aspects of formulating polymer concrete for pre-casting – the ease of manufacture, appearance, many properties and the cost are influenced decisively by these factors.

In general, therefore, pre-casting requires that considerable attention is given to these variables. Control of the reaction exotherm is also assisted by high aggregate loading, thus avoiding cracking, distortion or internal stresses.

Production

Two basic manufacturing techniques are available.

1. The resin binder and aggregate are mixed prior to placement in the mould which is then vibrated to achieve good compaction (a vacuum may also be applied to assist air release from the mixed resin prior to filling the moulds).
2. The dry aggregate is pre-packed in the mould and then saturated by introduction of the resin (vacuum techniques are used for this process).

Higher vibration energies are generally required compared to Portland cement concrete and are most reliably achieved with formwork vibration. The amplitude and frequency must be matched to the viscosity of the resin to achieve air release, consolidation and, therefore, high strength. Extended vibration times may be required to prevent segregation prior to gelation.

The second technique noted above is more expensive and difficult and, therefore, less common, although it may be preferred for special needs.

Staynes (1980 and 1981) in particular has investigated the vacuum impregnation technique for epoxy-based polymer concrete containing widely gap-graded, high strength aggregate. In addition to providing a very clean and controllable process, this method has other benefits for the production of consistently high quality polymer concrete, eg the complete elimination of moisture and air voids, the relative ease of attaining very high aggregate loading and control of aggregate interlocking.

With either production technique, considerable attention has to be given to the design of moulds and the use of effective release agents. The mould design for polyesters, vinyl esters and (meth)acrylates in particular must allow for the relatively high inherent cure shrinkage in order to achieve dimensional specifications.

The mould must be constructed of a durable, easily cleaned material which has good resistance to the liquid resin and a low coefficient of expansion. It must also be able to withstand the peak exothermic (or any imposed) temperature; control of the exotherm may be assisted by a high thermal conductivity.

Good sealing is essential and the design must provide: sufficient stiffness to avoid distortion during vibration; the elimination of trapped air and castings which require little, if any, finishing.

Where production is limited, the cost of designing and manufacturing moulds has a marked effect on the final cost of the pre-cast item. As with many similar processes, therefore, lower unit costs are favoured by high-volume manufacture of standardised (or customised) components.

Two options are available when the common pre-mixing technique is utilised, ie a batch or a continuous process. The latter demands much more sophisticated equipment (commercially available for a number of years (Kreis, 1981; 1987; 1990; 1992) and, consequently, a high capital investment.

However, this outlay provides certain benefits which become essential for long production runs or where large castings are produced:

- very precise mixing ratios can be achieved repeatedly both during one production run and over longer time periods (an important factor where consistent colour is essential)
- the time intervals for various steps can be optimised and accurately maintained
- high rates of mixing can be achieved readily, thus facilitating the use of active, fast-curing resin systems and increasing the rate of production (demoulding can be possible after only several minutes)
- production becomes less labour intensive
- costs associated with cleaning operations and wastage are minimised
- health and safety risks and environmental pollution are considerably reduced.

A detailed discussion of production methods has been given by Mikhailov et al (1992).

Current and future trends

It was noted earlier that growth in the use of pre-cast polymer concrete has taken place at a varying rate in different countries. The comparatively slow growth in the USA may be attributed, at least in part, to the high effort and investment which was devoted to the investigation of polymer impregnated concrete from the late 1960s onwards. However, as interest in this area peaked, in about the mid -1970s, and then declined somewhat in the 1980s, attention was increasingly focused on pre-cast polymer concrete (which in fact was used to produce building panels in the USA in as early as 1958).

A survey of likely future trends, discussed at a 1988 ACI Convention (Fowler, 1989) identified that pre-cast products, particularly utility components, would have the greatest impact in the area of polymer concrete. Additionally, it was predicted that pre-casting was most likely to be carried out by new companies, rather than by those with a background in reinforced composites or Portland cement concrete pre-casting. The following points also emerged from this survey:

- low-cost monomers/resins, obtained from re-cycled plastics, would become available and benefit the cost-effectiveness of pre-casting
- a much wider range of aggregates would be used, including materials made from scrap products; for example, in pre-cast machine bases
- new and improved equipment for pre-casting high quality components would be developed (also minimising health and safety risks)
- composite items, where thin sections of polymer concrete face a core of Portland cement concrete or a cellular polymer, offer considerable potential for building panels and other components.

In view of the primary general prediction given above, it is noteworthy that the 1991 ICPIC Workshop, held in San Francisco, contained a significant number of sessions dealing specifically with pre-casting.

An overview of the applications and potential growth areas for polymer concretes in Europe was given by Nutt in 1987 (see also Nutt, 1985). Significant changes over the intervening period are not obviously apparent in the UK. Consequently, the reasons advanced by Nutt for the very slow growth, particularly in comparison to Japan, may remain valid.

It was noted by Nutt that the disparity in cost between Portland cement and polymer concrete dominates commercial prospects for the latter. The description of such economic constraints as "apparent" would suggest that over-emphasis on the relative initial cost tends to control selection and specification. More specifically, for drainage and similar components, which are more easily handled and have relatively sophisticated designs compared to the alternatives, it was stated that:

> "The reasons quoted for a much greater switch to cast polymer concrete for public authority services in some other markets, eg Japan, have not yet found attractions in Europe; the facility to install or re-instate underground street services quickly and justify any extra cost burden by savings in the cost of traffic disruption is not the normal comprehension of those public servants who make the specification decisions in Europe."

Thus, it was suggested that pre-casters need to market the merits of polymer concrete components more aggressively and to demonstrate that associated and long-term cost benefits justify the initial expense. However, other factors may be particularly relevant to these products, eg a relative lack of investment in pre-casting and the inherited structure and attitudes of parts of the UK construction industry. Many factors in this area are likely to change as the industry becomes more influenced by overseas competition and adopts new practices such as "fast-track" construction.

As shown elsewhere in this report, a significant market penetration has been achieved over the past 30 years, or so, by other resin products which are now used routinely on civil engineering and other construction sites. Undoubtedly, this growth has been stimulated by the competitive market generated by the presence of numerous product suppliers, many of whom operate on an international basis.

However, for many of these on-site applications, a detailed analysis of lifetime cost benefits is not often required to justify their relatively high initial cost. While the need for durability is by no means ignored, there are usually compelling shorter-term attributes which lead to the acceptance of this high cost – for some applications there may, in fact, be very few, if any, viable alternatives. Thus, many uses are driven by the need for high chemical, abrasion or impact resistance, a fast rate of cure, high strength and adhesion, requirements which resin products uniquely satisfy.

In contrast, traditional alternatives to pre-cast polymer concrete are readily available. These usually have a lower initial cost and a reasonable service life. Consequently, it may be difficult to convince potential users that polymer concrete can yield longer-term cost benefits. However, the significant advantages associated with low weight may be accepted more readily, particularly for "fast-track" construction. A comparison can be drawn here with the use of resins for segmental construction. This innovative application was accepted early on in the UK because it was realised that, amongst other benefits, resin adhesives led to significant cost savings in the construction process itself.

The enormous market penetration of PVC components over the past 20 years, or so, demonstrates that polymer-based materials with a favourable cost/performance ratio do become accepted for routine building applications in UK., albeit more slowly than in other countries.

Furthermore, in civil engineering, the UK has demonstrated a willingness to lead in the development of new ideas. For example, a 63 m main span footbridge constructed entirely from polymer materials (filled GRP components, epoxy adhesives and Kevlar cable stays) was erected at Aberfeldy in Perthshire in 1992 (Richmond, 1993). This award-winning design (Plastics and Rubber Weekly, 1993) shows that high cost "new materials" are accepted (in the UK.) in demanding and innovative applications due to a combination of performance and significant cost savings in both the construction process and subsequent maintenance (a life to first maintenance of at least 30 years is claimed for the Aberfeldy Bridge).

On the basis of these examples, it may be anticipated that the established overseas acceptance of pre-cast polymer concrete may yet stimulate increased use in the UK.

3.13 MISCELLANEOUS APPLICATIONS

3.13.1 Sealing of joints

Expansion joints

Many structures require joints to provide continuity between building elements and to accommodate relative movement which would cause failure if otherwise restrained. Where these movements are fairly high (> approximately 10 per cent a typical joint width of $\geq$10 mm) and cyclic, for example due to thermal expansion/contraction, a conventional sealant is normally used in the joint (which is then referred to as an expansion joint). Such sealants range from bitumen and drying oil-based materials to highly elastomeric silicones, polyurethanes and polysulphides which may accommodate joint movements of up to ±30 per cent (see Appendix A1.4).

An essential feature of these sealants is that they undergo this type of movement without imposing undue stress on the bond to the substrate. This may be achieved with materials which have a high degree of plasticity, ie they stress relax rapidly compared to the rate of joint movement, with materials which have a very low modulus (typically 1–2 MN/m^2) or, very often, by a combination of these properties.

Products based on thermosetting resins, even those with a relatively low cross-link density which are often described as "highly flexible", are generally still too stiff for successful use in expansion joints. However, resin-based systems can be used in a related type of movement joint, as discussed below.

Contraction joints

The dimensions of concrete slabs commonly have to be limited to avoid cracking due to restrained cure shrinkage. Thus, flooring bay sizes are limited (or sawn joints which act as crack inducers are formed very shortly after concrete placement). The major part of the concrete shrinkage, which generally occurs during the initial 28 days, can thus be accommodated.

The resultant joints, generally referred to as "contraction joints", will continue to increase in width, but only to a very slight degree. Additionally, only minor expansion/contraction (due, for example, to thermal cycling) should occur at these joints as movement generated by the sub-structure should be allowed for in the main expansion joints. In many industrial situations, the sealant used in contraction joints has to fulfil the following functions, in addition to accommodating a relatively low level of movement:

- to provide good resistance against chemical attack from wet operations, spillage, and wet cleaning
- to provide a tough resilient seal against mechanical action and to rigidly support the joint arris against high compressive loads, eg from the wheels of fork-lift trucks.

Two-part epoxy-based sealants, often modified by urethanes, are commonly used in these joints as are some (meth)acrylate systems. As most joints are horizontal, these products are usually supplied with a pourable self-levelling consistency. In some circumstances, the use of a separate primer may also be recommended. Good performance, which can extend over many years, is dependent on two main factors.

1. The overall movement accommodation is generally limited to 5–10 per cent of the joint width. Installation should generally be carried out, therefore, after the initial concrete shrinkage has taken place. This may be in the final stages of construction and close to handover (however, at 10–20°C, sufficient cure for service is usually obtained within several days).
2. Good preparation and design of the joint is essential. Thus, de-bonding tapes or backing strips must be used so that the correct (recommended) width/depth geometry is obtained and adhesion is confined to the sides of the joint only (movement would be restricted by three-sided adhesion).

A product should be recommended by the supplier for a specific application of this type. The description "highly flexible" cannot alone be taken as a recommendation for use in joints which are subject to any movement. Similarly, the fact that polyurethane sealants are commonly used in expansion joints does not indicate that any "flexible polyurethane" is suitable; in this case, similar chemistry can be used to provide completely different physical properties.

Miscellaneous joints

In addition to contraction joints, such sealants may also be used in some construction joints and some expansion joints in tiled areas (in the latter case preformed composite joint fillers, eg metal/rubber strips, are perhaps more common). Epoxy-based compounds have also been used to seal the joints between pre-cast concrete segments in tunnels and subways.

Flexible polyester grouts have been used to seal the horizontal slots which carry electrical cables in concrete and asphalt pavements and to bed and seal airport lighting units. Tough resilient sealants which accept a limited degree of movement can be formed from these pourable fast-curing grouts.

3.13.2 Expansion joint nosings

The fragility of the joint arris on heavily trafficked floors was referred to in Section 3.13.1. Where concrete failure has occurred repair is often effected by forming a new nosing with a resin mortar.

Related nosings on stairways may also be strengthened by resin mortars or by resin-bonded metal strips or edging angles.

Steel edging angles, anchored or bolted into the concrete, have also been traditionally used to form the arris of the expansion joints in concrete bridge decks (where rapid wear is caused by traffic impact and vertical movement). However, permanent fastening can be difficult, as is repair or replacement following any disintegration of the supporting concrete.

To economically alleviate such problems, nosings formed in-situ from resin mortars were introduced on UK highway bridges in the early mid-1960s. These nosings were generally formed easily and rapidly with trowel-applied epoxy mortars; trials were also conducted with polyurethane and polyester-based products.

Although initially successful and extensively used up to the early 1970s, failures of new installations were increasingly observed in the form of hollowness/de-bonding and transverse cracks which often extended into the concrete substrate (Shaw, 1982; Mander, 1981). Extensive investigations revealed that certain features of the resin formulation, hitherto incompletely appreciated, were critical if those failures which could not be attributed to poor workmanship were to be avoided. These were:

- very active systems, designed for fast-cure at low temperatures, often have a very exothermic reaction, particularly at higher temperatures. Restrained cooling contraction can then lead to early failure by cracking or delamination. Later failure can also occur if such shrinkage gives rise to unrelieved stress concentration which is then exacerbated by differential thermal movement
- many of these reactive systems also tended to be too brittle with a low ultimate failure strain. At that time, the critical need for a relatively low modulus and high stress relaxation was not fully realised.

Such problems can be avoided by reducing the cure exotherm (less active systems; larger aggregate; carefully selected filler loadings) and by employing tougher, more flexible systems which show some degree of stress relaxation. Products having these properties were in fact used in the early, more successful nosings.

The use of polyesters and polyurethanes, which cure reliably at low temperatures, was limited in this application by their less reliable adhesion to damp surfaces and, in the former case, by the inherent cure shrinkage.

Resin-based nosings are now rarely used in the UK where bridge joints are generally formed from pre-fabricated steel/rubber units.

Loss of confidence in resin systems following failure, which in the above case was widely publicised, may well typify more widespread individual experiences. Thus, resin systems as a whole class can unjustly be labelled as "problem materials" on the basis of isolated failure which arises from very specific circumstances. In the case of nosings, some very reactive but unsuitable mortars were introduced in an effort to reduce material costs. In other cases, incorrect use may lead to failure.

While the possibility and consequences of failure will continue to vary from one application to another, sufficient knowledge of resin systems is now available to identify the likely problems connected with any use. Consequently, and as emphasised elsewhere in this report, an open discussion of potential problems is essential to avoid the general loss of confidence which can follow over-selling and unsatisfactory performance.

3.13.3 Waste encapsulation

The construction industry is directly concerned with containment and storage facilities for all manner of waste material. It also becomes involved with the treatment, eg solidification, and transportation of hazardous waste, particularly where Portland cement concrete or related materials are used to assist these operations. The encapsulation of low toxicity waste can also benefit the construction industry, in addition to society more generally, as this form of re-cycling can provide new types of low-cost composites which have many (potential) applications in building.

Increasingly over the past 10–20 years, a great deal of research and development work has been concerned with these areas as the management of industrial and municipal waste, including radio-active material, has become a major issue in many parts of the world.

In the context of the present report, it is noteworthy that several recent conferences in the USA, dealing specifically with construction, have included contributions on the use of reactive polymers for the treatment of waste (see Bibliography: ACI, 1989; ICPIC, 1991). Earlier work cited in these papers indicates that there has been activity in this area since the mid- to late-1970s. However, with only a few exceptions, the present status of reactive polymers in this application is summarised by "potential" rather than "routine and widespread". Consequently, the information provided here has been limited, although some indication of the potential is conveyed by the following examples:

- steel-fibre reinforced and polymer impregnated containers for the processing and disposal of radio-active waste (low and intermediate level) have been manufactured in Japan. These show certain benefits compared with steel drums and have been evaluated for deep sea or underground burial
- a (meth)acrylate and an emulsifiable polyester resin have been investigated in the USA as binders for the encapsulation of geothermal by-products containing toxic heavy metals. Standard polyester resins have also been investigated for the solidification of pharmaceutical and toxic chemical wastes
- glass-polymer composites, formed from unsaturated polyester resins and crushed waste glass, are claimed to be potentially competitive with reinforced concrete or vitrified clay for the manufacture of sewer and storm drain pipes. Other possible uses of this type of composite include tanks, building blocks, panels and "furniture" for public areas
- the ash given by waste incineration has been considered for use as a fine filler in reactive polymer systems.

3.13.4 The modification of OPC concrete

The addition of two-pack water-dispersible epoxy resins to conventional Portland cement mortars and concretes can lead to improvements in mechanical strength and the resistance to chemical attack and liquid penetration. The resin system also has a plasticising effect on the wet mix, thus assisting the use of low water/cement ratios to reduce drying shrinkage and porosity.

Such benefits are generally obtained at addition levels up to approximately 20 per cent weight resin/weight cement and can be particularly useful where relatively thin sections are applied (water loss by evaporation is also somewhat retarded by the epoxy resin, thus benefiting the cement hydration reactions).

Although references to this use of epoxies are available from at least the early 1970s (Hewlett and Shaw, 1977; and references therein) it would appear that the application has never become widely established. Undoubtedly, the enormous growth in the use of non-reactive polymer dispersions, eg SBR and acrylics (see Appendix A1.5) has contributed significantly to this lack of exploitation. These dispersions are easier to use than the epoxies and, in general, provide similar improvements at a lower cost. However, several publications from early to mid-1980s have claimed that modification by epoxies is viable, providing good resistance to chloride ingress and specific benefits for the repair of reinforced concrete by guniting (Christie et al, 1981; Popovics, 1985; Schorn, 1985).

4 Economic considerations

4.1 INTRODUCTION

Many of the generic components of resin-based products are derived from petroleum, usually by complex multi-stage processing. Conversion of these raw materials into formulated systems requires development work and additional factory-based operations. Initially, therefore, these products have a high cost compared to more traditional construction materials.

As shown in Section 3, reactive polymers can satisfy unique demands where a high performance is obtained from a relatively low usage, for example; resin injection, externally bonded reinforcement and anchoring. In these cases, use and the selection of individual products tends to be dominated by performance properties rather than by costs.

Much higher resin volumes are used for some applications where high performance is critical, for example; segmental construction and skid resistant surfacing. There, the general cost-effectiveness of reactive polymers is (or has become) self-evident. Lower cost products, most likely of the same generic type, may be considered as alternatives to established systems. However, as good long-term performance is so fundamental to such applications, considerable evidence may be required to demonstrate that it will be satisfactorily achieved before a lower cost product is seriously considered.

As noted in Table 4.1, the selection of particular resin-based products is not necessarily dominated by uniquely high performance properties as in the examples given above. Where a range of products, including non-resin materials, is known to provide performance above a certain level, detailed economic analysis may assist selection. Here then, a specific product might be favoured if reduced maintenance, and/or increased service-life, adequately compensates for a higher initial cost.

Table 4.1 *Reasons for using resin-based products versus cost analysis*

Demand	Comments regarding cost analysis
Urgent needs – solutions to problems occurring during/shortly after construction	Non-budgeted/unforeseen work - selection dominated by urgency. Insufficient information (and time) for detailed life cost comparisons.
Essential but less urgent need to remedy deterioration/failure Upgrading elements/structures	Comparison of whole life costs may be required for effective selection. May support use of reactive polymer rather than non-resin alternative.
Use in routine construction processes to achieve specific performance requirements	Whole life cost may be necessary to justify high initial resin cost. Life cost comparisons useful in assessing resin product options.
Alternatives to other materials for improved cost-effectiveness – reduced maintenance/improved service life	Whole life costing beneficial.
Use in innovative and/or prestigious construction	Selection dominated by performance.

Typical applications where such cost analysis may be of benefit include the protective coating of steel and concrete, the provision of floor toppings and the use of pre-cast components. In these areas, the long-term advantages of resin-based products (compared to other materials) have generally been promoted in qualitative terms. However, there is an increasing need to clearly evaluate their economic benefits as:

- a greater number of high performance alternatives now exists
- the economic viability of many industrial/commercial operations can be critically affected by loss of production/service during maintenance periods
- successful case histories do not necessarily confirm that the most cost-effective option was selected
- whole life costing is now becoming more routinely required in construction operations.

4.2 COST AND VALUE ASSESSMENT

4.2.1 Introduction

A floor topping is selected here as an example to illustrate one approach to an economic analysis. Within a new construction programme, a floor topping can be a critical programme item as it is one of the last activities to be completed before hand-over. However, the actual cost is normally a minor proportion of total construction costs. In contrast, flooring can represent a relatively major financial item within a maintenance budget. Replacement costs can be very significant due to the preparation work required. Disruption during repair or replacement can also be particularly important if critical activities cannot continue during the application and cure period, eg in factory production areas. Consequently, in addition to the more obvious factors, the cost-effectiveness of a floor topping can be affected by the following:

- the frequency of required maintenance, its ease and its contribution to extending the service life of a whole floor area
- the ease and cost of providing a new topping at the end of the service life.

Where it is known, from previous experience, that a certain type of topping is satisfactory, then the purchase costs of a limited number of similar proprietary options may be simply compared. In this case, it may be imprudent to select the cheapest option if there are any doubts concerning the performance claims, technical support and proven record.

In other cases, there can be a responsibility on designers and specifiers to select materials by a more systematic financial approach taking relative costs into account once the technical review has provided an initial short list of acceptable options.

An economic evaluation should consider:

- initial materials and labour costs
- maintenance, repair, replacement and disposal costs.

In this instance, costs need to be placed against relative technical and operational performance in order to make a comparison.

4.2.2 Initial cost

The initial cost of materials can vary significantly depending on:

- the country of manufacture (affecting labour costs, taxation)
- volume of production, amount of resin used in the formulation, general factory management and distribution efficiency, reinvestment in research, marketing costs and gross profit margin.

The contractor application costs can also vary, depending on such factors as:

- previous experience, protective measures to workforce or occupants, access, preparation, workforce skill level required
- number of applications of material, total area and the area that can be applied in one day
- method of application, curing and hardening times.

The above costs are not available to most purchasers and the market price of combined supply and application is obtained by seeking quotations from specialist applicators. Most manufacturers of high performance materials will have a list of approved applicators or at least local previous users.

An example of the first stage of a cost comparison taking into account the selection of resin finishes to improve floor performance is shown in Table 4.2. The costs (supply only) are for the UK with 1993 as the base year (see also Section 3.11).

Table 4.2 *The relative cost of floor finishes (Shaw, 1994)*

Type of finish		Cost (£/m^2)
1	Penetrating (solvented) two-pack epoxy	2–3
2	High build epoxy surface coating	4–7
3	Self-smoothing epoxy topping (1–2 mm thick)	10–16
4	Trowelled epoxy topping (6 mm thick)	18–24

This direct cost comparison is obviously of limited value as the expected performance of each material is significantly different. Two component reactive epoxies in solvents are able to penetrate both porous and dense floors and do not rely on adhesion. If a coating is considered as an alternative then the required impact or abrasion resistance becomes important, in addition to adhesion. When excellent chemical resistance coupled with high abrasion and impact resistance for heavy rolling loads is needed a trowel-applied resin mortar topping may be necessary. These differences will often mean that a simple cost comparison, as given above, is inadequate and the performance requirements have to be incorporated into a more complex matrix.

4.2.3 Maintenance, repair and replacement costs

Maintenance/repair costs and service life can be difficult to assess as comparative performance data are not readily available. It is necessary to discuss the intended use and the required lifetime with reputable companies who can often advise on suitable product selection if provided with adequate information from the specifier. It is an area where a specialist consultant can also provide good advice.

A checklist of factors, such as those given in Table 4.3, is useful when considering long-term performance. For example, clients may be advised that by changing one of their normal housekeeping operations they may realise substantial longer-term cost savings; changing to less aggressive cleaning fluids, placing certain operational restraints on the workforce or providing additional protection to the flooring in high wear areas, for example, may prolong the life of the floor and reduce maintenance.

Table 4.3 *Performance issues users should discuss with manufacturers/suppliers*

Checklist of performance issues	
1	Cleaning (method materials and frequency)
2	Maintenance to retain acceptable appearance/performance
3	Wearing/ageing characteristics and failure mechanisms, eg the ease of local treating or patching
4	Trigger for replacement, eg critical performance characteristics
5	Effectiveness of planned maintenance
6	Feasibility of overcoating/patching due to material compatibility
7	Ease of access
8	Possible NDT or sampling methods to assess residual life
9	Guarantees and support service from supplier
10	Flexibility of change in use or environment

There are also a number of general factors which users should consider before finalising any analysis and these are listed in Table 4.4.

Table 4.4 *Factors in the assessment of maintenance costs*

General factors	
1	Track record of the product in similar applications.
2	Are the references quoted by the supplier/applicator satisfactory and relevant to the specific application?
3	The cost of normal maintenance and repair work for the specific application.
4	How to ensure essential routine maintenance is completed to avoid premature need for replacement?
5	How can an adequate regular maintenance budget be obtained once installation is complete? Have future costs been assessed where proper maintenance is not carried out?
6	How to approach the problem if limited performance data is available in a meaningful form?
7	If whole life costs are to be calculated then what technique should be used to account for the timing of expenditure, eg is an organisation's agreed internal rate of return used to assess the "investment options"?
8	Is a sophisticated solution appropriate or too complex?

At some time, a flooring system will approach the end of its life and repair or replacement will be necessary. The time at which these costs need to be expended is a critical factor. The system's whole life costs may need to include replacement if the period of use exceeds the lifetime of the system.

Whole life cost-analysis and the modelling of service life performance of building and construction materials in general are still in the development stage. A particular problem concerns the scatter in observed performance which is very high due to the wide range of factors affecting ageing or damage. Technical data in this field is currently being developed by various research organisations.

The environmental and/or mechanical exposure of floor surfaces varies widely, as does the level of deterioration acceptable before replacement is deemed necessary (or the budget agreed to take the necessary action). For example, minor, acceptable wear and tear for many applications could be considered unacceptable in cases where damage could compromise hygiene standards or affect the operating performance of high-lift fork trucks.

The normal problem of the level and extent of historic economic and technical data required for a detailed analysis has the added complication for polymeric materials that formulations are modified to enhance performance.

It is suggested that expert advice is sought to assess the projected rate of ageing and time to replacement if a sophisticated whole life costing exercise is deemed to be necessary.

4.2.4 Guidance – lifetime classifications

Any meaningful discussion on lifetime will require the supplier of a product or advisor to know clearly the intended purpose and operational/environmental conditions. This is beyond the scope of this section. However, an approach to classifying life assessment is evolving in codes of practice, although relevant data is usually unavailable for the development of a quantitative model. Some guidance on the current approaches to lifetime performance can be found in the codes noted below.

BS 7543 defines service life as the period of time during which no excessive expenditure is required on operation, maintenance or repair. It is reasonable therefore to consider that non-routine costs only occur after the end of the service life.

BS 7543 has three categories of component life: replaceable, maintainable and long life. Polymeric materials fall into the first two categories with most floor finishes envisaged as being replaced during the building life. External cladding components are expected to last, with periodic maintenance, for the life of the building. However, the life of a building can vary from a minimum of ten years for temporary classrooms and office internal refurbishment, to over 60 years for health and educational buildings. Typical buildings will fall in the middle of these extremes with a minimum life of 30–60 years.

BS 8210 defines different maintenance levels once the durability limit (ie end of service life) has been reached. This standard also defines what should be understood by maintenance requirements and these are reproduced in Table 4.5.

Table 4.5 *Maintenance levels (other than daily and routine cleaning)*

Level	Description	Scope	Examples
1	Repair only	Maintenance restricted to restoring items to their original function after a failure.	Sealants
2	Scheduled maintenance plus repair	Maintenance work carried out to a predetermined interval of time, number of operations, regular cycles.	Five-yearly external joinery painting cycle. Five-yearly re-coating of roof membrane with solar reflective paint.
3	Condition based maintenance plus repair	Maintenance carried out as a result of knowledge of an item's condition. (The condition having been reported through a systematic inspection procedure).	Flooring wear

4.3 THEORETICAL CASE STUDY

4.3.1 Background

For many applications it is not necessary to distinguish between a wide range of options on economic grounds because choice is initially limited by service requirements. However, economic analysis may be appropriate and useful in the following two cases.

1. A comparison of proprietary products in the same broad performance category, eg resin-based floor toppings, possibly of different generic type.
2. Where the adequacy of a comparatively low-cost alternative is marginal in one respect or another but, overall, an acceptable balance of properties is provided initially. With service, the need for maintenance/repair is likely to increase in frequency and replacement/major upgrading is virtually inevitable after a relatively short, although acceptable, service-life.

The second case may be exemplified by a floor topping, used under physically demanding conditions, where some spillage of mildly aggressive (eg acidic) chemicals often takes place but washing-down is carried out irregularly. Here, a cementitious topping may be acceptable due to its low initial cost, although its ultimate performance is limited and fairly frequent patching is likely to be necessary.

Three alternatives with a progressively increasing initial cost (see Table 4.2) could be considered: a solvented, penetrating resin sealer, a high-build resin coating or a trowel-applied resin topping. Whether a cost benefit can be obtained, in comparison to the cementitious system, will be influenced by additional and variable factors, for example:

- the cost of initial surface preparation and application
- the frequency and ease of maintenance, including its effect on production (preparation required, health and safety, cure time)
- the degree to which the system is over-specified in terms of its ultimate performance/service life.

The first example given above has been selected for more detailed analysis. However, while the second example requires a greater range of cost estimates, the approach is similar in both cases.

Let us assume in the first example that the following criteria apply for a 2000 m^2 new flooring system for a medium duty industrial application. The principal requirements are for a flat, hardwearing, hygienic, acid resistant surface to be trafficked by plastic wheeled trolleys. There would be a major disruption to the operation if major repairs are required within the service life of the topping.

The specific technical requirements have been defined below against each key parameter adopting the code defined in CIRIA report R184 (Gatfield, 1997):

- structural base of permanent/in-situ concrete
- traffic conditions-plastic wheels (W2)
- skid resistance-high (E)
- abrasion resistance (W4)
- impact resistance-(IMP1)
- chemical resistance (M)
- level and flatness (3/SR1)

- ease of cleaning-high (H)
- acoustic-(u)
- anti-taint (T+NT)
- permeability (LVP/NWP)
- flammability (LF/MF)
- access to service (NS)
- appearance (U/S).

This combination falls into the categories where an epoxy, polyester, polyurethane or methacrylate resin system could be considered, albeit with varying confidence limits and anticipated service life periods.

4.3.2 Economic assessment

Installation and commissioning costs

The initial cost to a client of a flooring system will comprise design fees, site clearance and preliminary cost, specialist flooring contractor costs for materials and installation, site overhead and profit.

In this example, assume the main contractor was constructing an extension to a factory in 1995 and was not required to price the floor topping at the tender stage but allow for attendance and the provision of all support services. The construction programme shows that the flooring finish is on the critical path for hand over and the project quantity surveyor/main contractor have budget prices from specialist subcontractors for different systems that claim to comply with the performance specification. The main contractors' costs are common and, therefore, not included in the comparison given in Table 4.6.

Table 4.6 *Supply and application quotations for a floor finish*

Floor system type *	Service life (years)	Price quoted (£/m² laid)	Cost for 2000 m² (£)
Resin system A	10	33	66 000
Resin system B	15	40	80 000
Resin system C	20	46	92 000

* Systems A–C may represent self-smoothing or trowel applied floor toppings based on resins of a different generic type (see Section 4.3.1, Table 4.2 and, also, Section 3.11)

Lifetime costs

Future costs incorporate maintenance incurred by the owner after commissioning and hand over. They should also include cleaning, local repairing and replacement at the end of the lifecycle.

Selection has to be placed in the context of a given client's requirements and the risks that can be accepted. Thus, for operations where continuity of use is critical, an option which minimises the risk of disruption for repair/maintenance might be taken, even if this requires investment in a more costly product which could be considered over-specified; for example, a resin-based rather than a cementitious floor topping. In other cases, under specification may be viable if maintenance and repair requirements can be accepted while preserving the initial cost benefits.

The initial financial assessment needs a means to reflect the value of spending different sums of money now and in the future, whilst recognising the different lifetimes claimed. This is a common financial requirement and the Net Present Value (NPV) technique can be used to bring all future expenditure back to an equivalent present cost, recognising the cost of borrowing money (or a loss of income in the case of using funds available for investment).

The technique is explained in detail in most accountancy reference material. A simple explanation of the approach is that there is a choice of borrowing/ spending money now or delaying purchase and saving the repayment costs or investing the equivalent sum and spending it later after interest has been added. If there is a need to borrow the money at a rate of, say, 8 per cent, then delaying an expenditure of £100 by one year will save interest of £8. The net present value of spending £100 in one year is £100 x 0.92 = £92. Consequently, it is preferable to borrow £91 now rather than £100 in a year. Conversely, delay for a year is preferable to borrowing £93 now.

The technique is very sensitive to the annual cost of borrowing money that is assumed. This is normally variable and dependent on the financial status of the client and the sources of funds available. It does not take into account other financial restraints such as cash flow and overdraft limits, that could control a client's ability to select simply on this criteria. The certainty of future costs and the time of expenditure is also a significant variable.

Lifetime cost comparison

As can be seen from Table 4.7, the lifetime costs provide a ranking against the least net present value. However, the results are sensitive to the uncertainty of the input parameters. A slightly different judgement on the repair costs or lifetime could change the ranking. A very important point which is also business dependent is the residual value of the differing systems. If a factory process has an expected life of 20 years, there is no residual value in the floor topping after 20 years as the building could be totally refurbished or even demolished. Table 4.7 could be considered artificial, as it is unlikely that in the case of System C, a major floor repair would be instigated in the final year of operation. However, if there were hygiene concerns a major repair may be necessary.

Table 4.7 *Lifetime cost comparisons*

Floor system type	Service life	Initial cost (£)	Past service life repair costs* 10 years	15 years	20 years	Total (NPV at 20 years) (initial cost plus NPV of future cost ignoring residual value)**	Ranking
System A	10	66 000	33 000	66 000	–	100 075	2
System B	15	80 000	–	40 000	–	92 447	1
System C	20	92 000	–	–	46 000	101 434	3

* Assumed as 50 per cent of installation cost at the end of service life, total replacement at 1.5 x service life

** At an annual compound interest rate of 8 per cent, £100 borrowed in ten years is equivalent to £47.21 borrowed now. The figures for 15 and 20 years are £31.12 and £20.51 respectively.

A potentially major operation sensitive factor excluded from the above analysis is the disruption to a factory process if the floor requires repair or replacement. There can also be a particular problem with resin floors during repair where the volatile components can contaminate food.

The above analysis also ignores routine cleaning and maintenance costs as these are normally common or insignificant in the purchase decision.

Due to the uncertainties involved in a NPV financial analysis for this kind of comparison, it has to be treated as a technique which complements the conventional assessment of performance parameters in the selection process. It does not provide a definitive solution or an independent means of arriving at a decision. Hence scientific and engineering judgement must be included.

It is essential that these options are fully discussed with the client and consequently there is a need for the realistic assessment of both the limitations and the benefits of products.

Appendix 1: Related materials

A1.1 FIBRE-REINFORCED COMPOSITES

The enhancement of properties through fibre reinforcement is a key factor in the use of plastics for many engineering applications and is one that has led to high volume usage. Both thermoplastics and thermosetting resins benefit from fibre reinforcement, although the latter are most common for construction applications.

Given the high performance applications of non-reinforced thermosets discussed earlier, the need for further property enhancement could be surprising. However, in a broader context, Section 3 deals mainly with speciality applications, where the optimum use is made of certain characteristics. For large but lightweight rigid components with a low section thickness, whether flat sheets/panels or more complex shapes, the deficiencies which are largely "designed out" by the nature of the earlier applications are overcome by the incorporation of high-tensile strength fibres. Broadly, relatively brittle materials are thus made significantly stronger (although not necessarily appreciably stiffer). Cracks are arrested or deflected by the fibres. Consequently, fracture toughness can be dramatically increased, resulting in a substantial improvement in tensile and impact strength and a very high strength/weight ratio (at least twice that of mild steel).

Glass fibres are the principal form of reinforcement used for plastics because they offer a good combination of strength, stiffness and price. Carbon fibres and those from aromatic amides can give improved strength and stiffness but these materials are expensive; hybrid glass/carbon fibre systems provide a good balance of properties at a more acceptable cost.

Glass (fibre) reinforced plastic(s) (GRP) also have the following attributes:

- ease of mouldability and availability in many variations of colour and finish/texture
- impermeability and excellent water/chemical/environmental resistance.

By far the greatest proportion of thermoset GRP products used in construction are made from unsaturated polyester resins, although phenolic and (meth)acrylic resins are also used; the former due to their good inherent fire resistance and reduced smoke emission and the latter due to their excellent weathering properties. Epoxy and vinyl ester resins are used to a lesser extent due to their higher cost although, for some special applications, advantage may be taken of their potential for improved strength, toughness and chemical resistance.

A borosilicate glass of very low alkali content (E glass) is predominantly used in GRP in a variety of forms – most commonly as chopped strand mat, but also as continuous rovings which are usually chopped in the moulding (laminating) process, woven rovings, woven cloth and continuous filament matt. For good durability, it is essential that adhesion between the fibres and the resin matrix is maintained.

For GRP construction products, the original and simplest manufacturing technique, contact moulding (also known as hand laminating or hand/wet lay-up) is still widely employed. Spray moulding and a number of more mechanised techniques can also be used. For many applications, an initial (external) layer of unreinforced resin (the "gel coat") is used to enhance appearance, weathering and wear characteristics.

Post-curing at an elevated temperature is usually desirable and particularly necessary to maximise water/chemical resistance and performance at high temperatures.

While thermosetting GRPs have good strength, they generally have a relatively low stiffness, particularly compared to mild steel or aluminium alloys. Appropriate care is necessary in design, especially where there is a risk of buckling failure. The following methods are employed to increase the stiffness of GRP laminates:

- "shaping for stiffness", ie using folded or curvilinear forms
- increasing the glass fibre/resin ratio
- using woven rovings or cloths as reinforcement or introducing other fibres transversely to the normal laminar fibre orientation
- bonding to a core material – full, effective adhesion must be obtained and consideration must be given to differences in the thermal coefficients of expansion.

A very wide variety of core materials are used with GRP laminates, including polyurethane and phenolic foams. In addition to increasing stiffness and strength, core materials reduce installed costs and improve such characteristics as thermal insulation, sound attenuation, fire resistance, ruggedness and blast resistance.

The economic production of GRPs, together with the design freedom and beneficial properties which they provide, has led to increasing use in building and civil engineering applications over the past 40 years. Typical applications, amongst many others, include:

- architectural forms and features, eg domes, arches, fascias, mullions, canopies, cornices
- external cladding panels, roof panels, translucent profiled and flat sheeting, columns and balconies/balustrades
- modular building and kiosk systems
- water storage tanks, cisterns, septic tanks and other water/sewage/effluent control products
- pipes
- concrete formwork (for complex shapes/intricate detail and, also, as permanent formwork)
- site-applied GRP linings/membranes.

More recently, interest in GRP laminates for more highly stressed, "structural" applications has increased rapidly. Such uses have been restricted by a lack of available design information, methodology and history of performance in service. Consequently, the need evolved for a more formal but practical code addressing all aspects of design, construction, testing and quality control. Such a code has recently been produced by a European group with financial support from the EUREKA initiative (Clarke, 1996). It is anticipated that this EUROCOMP Draft Design Code could eventually form the basis of a Eurocode (Quinn and Godfrey, 1991).

The completion, in October 1992, of the world's first structural GRP footbridge, spanning the River Tay at Aberfeldy Golf Club in Scotland, illustrated the advances that could be made in construction with these materials (Richmond, 1993). This potential was extended, in 1994, with the opening of a road bridge, the Bonds Mill Lift Bridge, on the Stroudwater Canal near Gloucester (Anon, 1994; Nuttall, 1994).

Apart from laminates, GRP is also being considered more (and used) for concrete reinforcement, particularly in aggressive environments where steel corrosion is a major concern. Replacements for steel in bar, grid and tendon form are available.

A1.2 FORMALDEHYDE-BASED POLYMERS

As shown earlier (Figure 2.1), thermosetting polymers are classified as addition polymers (of main concern in this report) or condensation polymers. The latter most commonly utilise formaldehyde as one of the reactive monomers. Both types have a cross-linked network structure. However, whereas addition polymers can be formed under normal ambient conditions, high processing temperatures, often well above 100°C, are generally required for the production of condensation polymers. Additionally, high pressures and specialist equipment are usually required for fabrication, thus preventing the formation of voids by the volatile molecules evolved during the cure reactions. The most common condensation thermosets are:

- phenolic polymers derived from phenol or related compounds such as resorcinol and an aldehyde, usually formaldehyde or furfural
- amino polymers obtained from formaldehyde and urea, melamine or similar monomers.

These polymers are formed in several stages, the first being the formation of the resin itself, often in the form of a moulding powder; the last stage, the cure or cross-linking reaction, is usually conducted during the fabrication process at temperatures between 150–200°C. Fillers, incorporated at an intermediate stage, include wood flour, cellulose fibres, minerals, rubber and glass fibres. They both reduce cost and selectively improve properties. Both the cure temperature and filler type affect the mechanical properties and other characteristics such as the chemical and fire resistance.

Phenol-formaldehyde polymers, only available in dark colours, are strong and hard, but brittle, with excellent creep resistance and good electrical insulation properties. Surface hardness can be affected by hot and damp conditions, although the more general weathering properties are excellent. They are susceptible to attack by alkalis and oxidising agents; the phenol-furfural polymers give improved chemical resistance. The applications of phenolic resins are summarised:

- tile bedding and grouting
- moulding resins – the products have outstanding heat resistance (to approximately 200°C) and dimensional stability and very good dielectric properties. They can withstand very high temperatures for brief periods and then carbonise, maintaining a protective thermal barrier
- laminating resins, eg for worktops, wall coverings, printed circuit boards
- bonding resins, eg for brake linings, abrasive wheels and foundry moulds
- casting resins – giving products with excellent temperature and flammability resistance and good machinability
- coatings and adhesives – particularly as wood adhesives, eg for plywood (resorcinol-based wood adhesives cure at room temperature).

Compared with the phenolics, amino resins have the benefit of being clear and colourless and their tensile strength and hardness are generally better. The melamine resins have superior hardness, heat resistance and moisture resistance to the urea-based resins; when mineral filled they can be used at continuous service temperatures approaching 150°C. They are also self-extinguishing, good electrical insulators and have good abrasion/ scratch resistance.

Their applications are summarised as follows:

- moulding resins – widely used for domestic items, appliance housings
- laminating resins – decorative laminates for counter, table and worktops
- adhesives, particularly for plywood bonding
- enamels for domestic appliances.

A1.3 CELLULAR POLYMERS (FOAMS)

As most thermoplastic and thermosetting polymers (and rubbers) can be foamed, a large variety of cellular products (expanded plastics) are potentially available but relatively few generic types have been exploited commercially. Those that are produced range from hard, rigid materials at one extreme to very soft, flexible materials at the other and include products which are classed as structural foams for load-bearing applications (fibre reinforcement can be incorporated in some cellular polymers). An enormous variety of industrial and domestic uses contributes to a market size which, in Western Europe, has been estimated to be of the order of 1.4 million tonnes in 1990.

Common cellular thermoplastics are based on PVC, polystyrene, polyethylene and related polymers, polycarbonates and ABS polymers. Polyurethane is by far the most versatile and widely used thermosetting polymer but urea-formaldehyde and phenolic foams are also very common. PVC and phenolic foams are particularly attractive because they do not support combustion, possessing "natural" flame-retardant properties.

Cellular polymers are two-phase materials, consisting of an inert gas dispersed in the solid polymer as relatively small bubbles which, in principle, may be discrete (closed cell foams) or connected so that both phases are continuous (open cell foams). In practice, the two cellular morphologies often co-exist, the volume fraction of closed cells having a considerable influence on the mechanical properties of the foam. The ease with which the gas phase can move through the polymer matrix also governs the mechanical and other physical properties.

Cellular polymers are commonly classified as low, medium or high density. The distinction is not exact or strictly defined but the following figures give an approximate guide: 10–50 kg/m^3, 50–350 kg/m^3 and 350–900 kg/m^3.

The expansion of the polymer (and, therefore, the gas in the cells) originates with an additive known as a blowing agent which can operate either physically or chemically. In the former case, the blowing agent undergoes a phase change, for example: heat changes a volatile organic liquid into the gaseous state or a gas dissolved by compression (such as air or nitrogen) is liberated by release of the internal pressure. Chemical blowing agents are usually mineral or organic compounds with good thermal stability at ambient temperatures. At a well-defined elevated temperature, these materials decompose with the liberation of large volumes of gas, commonly nitrogen or carbon dioxide; in some cases, decomposition is chemically induced by the addition of an activator.

When polymerisation or cross-linking occurs concurrently with gas evolution the two processes must be correctly balanced (as in the case of polyurethanes which can be foamed in place at room temperature). If gas evolution is too rapid, then the foam will collapse, after an initial rise, as the matrix will be too weak to retain the gas; in the opposite case, the foam does not rise adequately.

Cellular polymers are conventionally classified as flexible, semi-rigid or rigid with a further sub-division of the latter according to use, ie non-load bearing for applications such as thermal insulation or load bearing where high stiffness, strength and impact resistance are essential. By selecting the appropriate components, the degree of cross-linking of polyurethanes can be widely varied thus giving either rigid or flexible foams, both of which are commercially important.

A "structural foam", which can be obtained from most thermoplastic and thermosetting polymers, has a full-density skin and a cellular core, ie a sandwich configuration with a favourable strength/stiffness-to-weight ratio.

Syntactic foams, which are used as core materials in composite sandwich construction and for thermal insulation, utilise pre-formed hollow bubbles of glass, ceramics or polymers in place of a blowing agent. Epoxy and polyester resins are common matrix materials.

Properties such as strength, stiffness, creep and temperature resistance can be improved by reinforcing the cell walls of a foam using, for example, short glass fibres which are introduced prior to blowing. These reinforced foams are used with various types of facing sheets, eg aluminium foil, as building and insulation laminates.

Some properties and applications of a number of cellular polymers are given below.

1. **Polystyrene**: rigid, closed cell foam with a density ranging from 16–480 (mostly 16–80) kg/m^3; good moisture but poor u.v. resistance (may absorb significant water in wet situations where the temperature varies); useable to about 70°C; flammable; used in many applications requiring thermal insulation (roofs, walls above/below ground) and as a core material for sandwich panels; used as "shaped formwork" to produce holes/channels in concrete; used in bead form in lightweight, insulating concrete; pre-expanded beads are also fused to give board/slab forms; used as permanent shuttering for concrete.
2. **PVC**: the closed cell rigid form (density 25–125 kg/m^3) has good all-round properties and, in particular, a low flammability; used in sandwich panel construction.
3. **Polyurethane**: usually has a density of 30–35 kg/m^3 in the closed cell, rigid form; good thermal resistance (to 120°C); low water vapour permeability and thermal conductivity, although the latter often increases with age; poor u.v. resistance and behaviour in fire (although the toxicity of the pyrolysis products is no worse than for many other materials) – flame retardant grades are available; has a multitude of insulation uses and is used as a core material for sandwich panels; has the considerable benefit of being easily foamed in-situ and thus can be used for spraying over large surface areas or for filling irregular cavities and voids, eg at service entrance points in walls.
4. **Phenolics**: good thermal insulation (closed cell) and very low flammability and smoke generation; good resistance to temperature (150°C), chemicals and water uptake/transmission.
5. **Urea-formaldehyde**: very low thermal conductivity with little variation for foam density of 10–30 kg/m^3; negligible strength and, therefore, usually foamed on site using portable equipment; can cause metal discoloration (copper, aluminium) or corrosion (zinc, steel); high water absorption; deteriorates continuously at a moderate rate which is accelerated by temperature and high humidity – toxic formaldehyde is generated; supports fungal growth whose spores can cause allergic reactions.

A1.4 JOINT SEALANTS

Non-rigid joints, which have to be sealed for a variety of reasons, occur or need to be formed in the majority of building and civil engineering structures, in addition to roads and airfield pavements.

In buildings in particular, marked changes in facade construction methods over the past 40 years or so have led to increasing technical demands in this area. Thick, load-bearing masonry has been replaced by curtain walling systems in which pre-fabricated relatively lightweight panels are fixed to a structural steel or concrete frame, thus forming a non-load-bearing external "envelope".

Such systems assist in the construction of high-rise structures, with an associated increase in joint lengths (which may total close to 100 km on some buildings). The need for a greater number of joints also arises, due to the multi-component nature of such facades and their inherent flexibility.

As a consequence of these developments, greater performance demands were placed on the materials used to seal the joints – greater levels of movement needed to be tolerated if the number of joints was not to become excessive; good adhesion to a greater variety of surfaces was required; and cohesive/adhesive integrity needed to be maintained for long periods due to the difficulty and high cost of replacement. Thus, the older, essentially non-hardening caulks and mastics, often based on bitumen, butyl rubber or naturally occurring resins, were no longer adequate and new wet-applied materials with a higher and more closely defined performance evolved. These were invariably based on synthetic polymers and became known conventionally as "sealants". In addition, an increasing range of preformed synthetic tapes and gaskets were developed.

Joints occur, or can be formed, in various configurations (butt, shear, fillet) and may be subjected to various types and degrees of movement, ie repeated opening/closing (expansion or movement joints), only one mode of movement (compression and contraction joints) or very little movement (some construction joints). Joints also accommodate deviations from design dimensions. The function and design of joints is fully discussed in BS 6093 (1993), *Code of practice for design of joints and jointing in building construction* (other Standards are available for more specific areas of use, eg roads, glazing). The main functions and requirements of building sealants can be summarised as follows:

- to maintain a weather-tight barrier
- tolerance of site conditions during application and cure
- to bond reliably to appropriate substrates
- accommodation of complex diurnal/seasonal movement
- cohesive and adhesive durability under various climatic conditions
- acceptable appearance.

The use of relatively rigid thermosetting sealants in flooring contraction joints was discussed in Section 3.13.1. For accommodation of higher levels and rates of movement, thermoplastic and elastomeric sealants are employed.

The predominant thermoplastic sealants are based on acrylic polymers, used either in a latex form (mainly for internal service) or as solvented systems.

Sealants with greater elasticity are based on polysulphides, polyurethanes and silicones. These sealants can accommodate greater movement levels (typically a total joint movement in the range 25–75 per cent, based on the minimum width of a properly designed butt joint) and generally have a useful life expectancy of at least 20–25 years. All three types are available as two-part systems or as single-pack moisture cured materials; in the case of polysulphides and silicones, two-pack and single-pack products, respectively, are most common. The single component sealants offer greater convenience for application but, in general, the two-pack systems have a greater rate of cure and, thus, can accommodate high levels of movement earlier.

In order to achieve satisfactory performance, it is most essential that:

- the joints are correctly designed, ie number, location, type, dimensions; in particular, joints must be designed with a minimum width (with an allowance for construction tolerances) if a given level of movement is to be accommodated by a selected sealant
- the sealant is appropriate for the estimated range and rate of movement and the service environment
- the sealant is properly applied, with the correct width/depth ratio, to well-prepared surfaces; primers must be used where recommended and must be appropriate for the sealant/substrate combination.

Insulating glass provides an additional large market, amongst others, for sealants – here the sealant is used as an adhesive for holding the panels together and as a seal against the ingress of water vapour. A further notable use of relatively high modulus silicone sealants is in structural glazing applications where the sealant acts as an adhesive between the exterior glass panels and the supporting structure. The resulting "flush-glazed" facade is now a common feature of many cities around the world.

A1.5 POLYMER CEMENT CONCRETE

The modification of Portland cement concrete/mortar and other cementitious materials, using aqueous polymer dispersions has been referred to in other sections of this report as such use is very common in repair and flooring compounds (and, in the USA, in bridge-deck overlays).

Relatively few synthetic polymers are soluble in water. However, certain types can be produced as stable suspensions of solid polymer particles (typically with a mean size of 0.1–1 μm) in an aqueous medium. On drying, these dispersions give a coherent film via coalescence of the particles. The dispersions are also referred to as "lattices" or "latexes" (and, incorrectly, as emulsions; see Glossary) and are typically supplied by the manufacturer with a solids content of 40–70 per cent. Viscosities are often very low and certainly lower than would be the case for a polymer solution with the same level of solids.

Following polymerisation, the dispersion is compounded so that it is suited to formulation for a particular type of end-use, for example: an adhesive, paint/surface coating, sealant, an asphalt/bitumen modifier or as a modifier for cementitious compositions.

Although the milky white polymer dispersion is a seemingly simple liquid, the polymerisation and compounding processes are very complex. At the first stage, there are at least six or seven main variables which have to be optimised in order to obtain the dispersion properties required; for compounding, there are more than a dozen ingredients which need to be considered.

The benefit of these variables is that they facilitate matching the dispersion characteristics to a given end-use. They also lead to an overlap in the properties obtainable from different polymer types, suggesting that performance specifications should be used where possible rather than the prescription of a certain generic type.

Dispersions of elastomeric/thermoplastic polymers have been used to modify cementitious compositions since the 1920s when it was realised that natural rubber latex would improve the tensile/flexural strength, toughness, chemical resistance and adhesive properties of relatively thin (<50 mm) floor toppings, mortars and levelling/sound damping screeds used on the steel decks of ships.

Polyvinyl acetate dispersions (PVA) have been used since the 1950s as bonding agents for gypsum plasters, particularly where the substrate is normally unreceptive. However, while excellent properties are maintained in dry conditions, deterioration occurs in the presence of water (modified PVAs with improved water resistance are available but these may be decomposed by the alkalis in cement).

Later developments led to a greater variety of dispersions possessing the benefits of natural rubber latex and PVA but with less disadvantages. For use in cementitious systems, three generic types are generally found to be most useful and are widely used, ie styrene-butadiene rubber (SBR), polyacrylic esters (PAE, also known as pure acrylics as they are copolymers of (meth)acrylates) and styrene-acrylics (SA). Besides gypsum plaster and ordinary Portland cement, dispersion types/grades are available for high alumina, sulphate resisting and other cement types.

Certain types of dispersion (for example. ethylene-vinyl acetate) can be spray-dried to form powders which re-disperse easily in water, particularly under alkaline conditions; these powders are particularly useful for the production of single-pack cementitious systems. However, while generally satisfactory properties are obtained from these re-dispersible polymers, it is often recommended that an aqueous dispersion should be used where the maximum performance is required under wet conditions.

The relative merits of the three main dispersion types noted above (SBR, PAE and SA) have been the subject of considerable debate and, consequently, the reader is referred to the Bibliography for comparative information. The more general benefits of polymer dispersions in cementitious systems are summarised below (an appropriate dosage level and mix design and correct methods of working are assumed, the latter differing in a number of respects from unmodified mixes):

- the water/cement ratio can be lowered as the polymer contributes to workability; the drying shrinkage is then lowered (and other properties improved) – a particular benefit for thin layers as the risk of cracking is reduced
- significant improvements are obtained in tensile/flexural strength, strain capacity/ toughness and impact/abrasion resistance; good compressive strength is maintained
- excellent frost resistance, enhanced chemical resistance and lower permeability to gases, vapours and liquids; increased resistance to carbonation; good protection of steel against corrosion
- significantly improved adhesion to a wide variety of substrates.

The improvements in mechanical and barrier properties are considered to arise from the formation of continuous polymer films and strands within the cementitious matrix. These strands both bridge and partially block the normal internal voids thus reducing crack propagation and permeability.

Overall, the improved performance (and proven durability) of modified mixes offsets their increased cost (particularly where thinner sections can also be used). Consequently, dispersions, supplied either alone or as part of a fully formulated proprietary product, are widely used in such applications as:

- bonding agents and adhesives for renders, brick slips and tiles
- floor screeds, toppings and overlays
- mortars for repair, bedding, bricklaying and rendering
- cementitious coatings for the protection of concrete and steel, for levelling (fairing coats) and for waterproofing
- pre-cast products, eg tiles.

Appendix 2: Contents of Volumes 1, 2 and 3

A2.1 VOLUME 1: THE MATERIALS AND THEIR PRACTICAL APPLICATIONS (CIRIA C537)

A2.2 VOLUME 2: SPECIFICATION AND USE OF THE MATERIALS (CIRIA PR78)

A2.3 VOLUME 3: MATERIALS TECHNOLOGY (CIRIA PR79)

Glossary

ABS POLYMER	A terpolymer formed from acrylontrile, butadiene and styrene monomers.
ACCELERATOR	Any substance that increases the rate of a chemical reaction such as the curing or cross-linking (hardening) of a resin system; for polyester and similar resins it refers more specifically to an additive (also called a promotor) that accelerates the action of the peroxide catalyst.
ACRYLATE POLYMERS	See "acrylic polymers".
ACRYLIC POLYMERS	A collective term for an extremely diverse range of polymers and copolymers based upon acrylic and/or methacrylic acid and their derivatives. Polymers based on the ester derivatives of these acids provide a sub-group referred to as acrylates/methacrylates. Polymers within this range include thermoplastic, thermosetting and elastomeric materials.
ACTIVATOR	A substance added in small quantity to promote activity; peroxide catalysts added to polyester and similar resins are sometimes called activators (or initiators).
ADDITION POLYMER	A polymer formed by the combination of monomers (or the cross-linking of pre-polymers) without the formation of by-products, eg the splitting off of water or other simple molecules; thus, the structural unit of the polymer has the same molecular formula as the monomer itself.
ADDITIVE	Any substance added in small quantities to improve or modify one or more specific properties (thus a filler, pigment or solvent added in relatively large amounts is not an additive).
ADDUCT	A modified form of curing agent for epoxy resins (obtained by pre-reacting a lower molecular weight amine curing agent with a relatively small amount of resin).
ADHESION PROMOTER	A chemical additive designed to react with both the continuous polymer phase and a particulate or fibrous phase (or a substrate) and thus to promote a strong and possibly more durable interfacial bond (also known as a coupling agent).
AGGREGATE	Solid mineral-based material such as sand or gravel; fine aggregates (approximately 1–5 mm) are commonly incorporated in resin-based products and are often included within the collective term "filler".

ALIPHATIC	An organic compound consisting of straight or branched chains of carbon atoms.
AROMATIC	An organic compound containing one or more benzene rings.
BASE RESIN	The synthetic resin supplied by the manufacturer to the formulator and which may already contain some modifier(s).
BINDER	The continuous solid phase or matrix (generally the polymer) holding the particulate or fibrous phase in a composite material.
BISPHENOL-A (or -F)	Derivatives of phenol used in the manufacture of common epoxy resins.
BLOWING AGENT	A compounding ingredient used to produce gas by chemical/thermal action in the manufacture of cellular polymers (foams).
BONDING AID	A term often applied to systems used to adhesively bond new to old concrete, eg some epoxy resin products and polymer-modified cement slurries.
CARBON FIBRE	Fibres produced by the pyrolysis of organic precursor fibres in an inert atmosphere. Used to meet specific high performance requirements.
CATALYST	A substance that increases the rate of a chemical reaction (in strict terms the catalyst itself should remain chemically unchanged); in the context of this report the term is applied specifically to substances, such as peroxides, which induce cross-linking of polyester and similar resins (also referred to as an initiator or activator).
CE MARKING	The means of signifying that a product complies with the "essential requirements" of a European Directive, eg the Construction Products Directive.
CHAIN	A chemical structure in which each atom is joined to the next to form a chain or line of atoms; chains may be straight or branched.
CHALKING	The formation of a friable, powdery layer on the surface of a coating caused by disintegration of the binder surface by weathering.
COLD-CURING RESIN SYSTEM	One which hardens without the external application of heat; the type most commonly used in construction work.
COMPOSITE	A solid product consisting of two or more distinct phases, including a binding material (matrix) and a particulate or fibrous material; more specifically, reinforcing material of high strength/modulus/aspect ratio encapsulated by and acting in combination with a polymeric matrix.

CONDENSATION POLYMER	A polymer formed by the reaction of (generally) at least two monomer types and with, in most cases, the formation of a low molecular weight by-product, eg water; the structural unit thus contains fewer atoms than the monomers. May be viewed as a specific type of "copolymer".
CONTINUOUS FILAMENT	A yarn or strand in which the individual filaments are Substantially the same length as the strand.
COPOLYMER	A polymer formed from two types of monomer.
COUPLING AGENT	See "adhesion promoter".
COVALENT BOND	A specific type of chemical bond formed between atoms by shared electrons (an ionic bond is formed by the attraction of opposite charges on ions).
CREEP RUPTURE	Fracture of a material under creep conditions at a stress which is lower than the ultimate strength given under rapid loading.
CROSS-LINK DENSITY	A measure of the number of cross-links per unit volume of polymer – determined by the number of reactive groups leading to cross-linking and their distance apart. Has a major effect on many properties. Higher cross-link densities are generally associated with increasing rigidity and its retention at elevated temperatures.
CROSS-LINKING	The formation of a three-dimensional polymer (a network structure) from linear or branched polymers by means of interchain coupling reactions. Cross-linked polymers are generally infusible and insoluble and, therefore, cannot be remoulded.
CROSS-LINKING AGENT	A compound that combines with a reactive pre-polymer to give a three-dimensional network, eg styrene for unsaturated polyesters or amines for epoxy resins.
CURE	To change the properties of a system into a more stable condition and to develop the desired service properties – generally cross-linking by chemical reaction but also drying by loss of a solvent.
CURE TIME	The time required under specific conditions to reach a specified property level.
CURING AGENT	Generally a "cross-linking agent" but most commonly applied to amines and other compounds which are added to epoxy resins immediately before use to provide cross-linking (the different use of this term in concrete technology should be noted).
CYCLO-ALIPHATIC	An organic compound containing carbon atoms combined in any form of ring structure other than that of benzene rings.

DEGREE OF CURE (OR CROSS-LINKING)	The extent to which potential cross-linking reactions have progressed or to which ultimately attainable properties have been achieved.
DILUENT	A liquid additive used to reduce viscosity and to modify other properties of a resin system (both before and/or after cure); reactive diluents become chemically bound to/within the cross-linked network; non-reactive diluents are physically incorporated within the cured resin and may be lost by diffusion/ evaporation/leaching; most commonly applied to specific additives used in epoxy resins.
DISPERSION	A two-phase system in which a disperse phase is permanently distributed as small particles throughout a continuous phase, eg a stable suspension of very fine particles in a liquid medium.
ELASTOMER	A polymeric material that, under service conditions, returns rapidly to approximately its initial dimensions and shape after release of a stress that has led to substantial deformation.
EMULSION	A two-phase mixture of normally immiscible liquids which is stabilised by certain additives; one liquid is normally dispersed as very fine droplets in the other, eg an oil in water emulsion. Note: this term is often applied erroneously to stable emulsion-like dispersions of a solid polymer in water, eg polyvinyl acetate "emulsions"; similarly, the conventional use of the term "emulsion paint" is strictly incorrect.
EPOXY RESINS	Both the relatively low molecular weight resin precursors containing epoxy (epoxide or oxirane) groups and the derived cross-linked polymers are referred to as epoxies or epoxides.
ESTER	An organic compound formed when an alcohol reacts with an organic (carboxylic) acid giving the characteristic ester group.
EXOTHERM	The evolution of heat during a chemical reaction or change of state, eg curing reactions of resin systems.
EXTENDER	A liquid material used to augment the binder, generally to reduce cost but also to modify properties, eg a non-reactive hydrocarbon resin; also refers to inorganic materials in powder form used as pigments/ fillers in coatings.
FAIRING COAT (LEVELLING COAT)	A material or product applied (usually to a concrete surface) to give a smooth surface prior to coating.
FILLER	Strictly a very fine powder (<1 mm) addition; also used to denote granular (fine-coarse) aggregates, some fibrous and other solid materials added to resin systems, generally to reduce cost but also to modify properties; fillers are always insoluble in the resin.

FILLER/BINDER RATIO	Generally the weight ratio between solid (insoluble) and non-volatile liquid components in a resin system (usually applied only to solvent-free products).
FLASH POINT	The minimum temperature of a liquid at which the vapour given off is sufficient to form a flammable mixture with air, under specified test conditions (flash point depends on the method of test).
FLEXIBILISER	A diluent or extender incorporated into a resin system specifically to lower the modulus of the cured product.
FORMULATION	The fully detailed composition (or mix design) of a resin-based product.
FREE RADICAL	A (normally) very reactive chemical species which (in contrast to an ion) is uncharged.
FRP	See "GRP".
FULL CURE	Conventionally (for resin systems cured at normal temperatures) a sufficiently high degree of cross-linking for the complete development of service properties (see also "post-curing").
FUNCTIONAL GROUP	An atom or group of atoms which gives an organic compound its characteristic chemical reactivity, eg an epoxy, amino, hydroxy, vinyl, or isocyanate group.
FUNCTIONALITY	The number of reactive groups within a molecule.
GEL COAT	A resin used in moulding processes to provide an improved surface on a composite; the first resin applied to the mould after the release agent.
GEL POINT	The stage of cure at which a liquid resin system begins to exhibit a degree of cohesion and elasticity (related terms: gelation, gel time).
GLASS TRANSITION	The reversible change in a polymer from (or to) a hard and relatively brittle state to (or from) a viscous or rubbery state – a transition associated with a change of molecular motion due to a change of temperature (may also be referred to as a second order or a rubber transition). Many properties undergo a marked change with this transition.
GLASS TRANSITION TEMPERATURE/POINT (Tg)	The approximate mid-point of the (usually relatively narrow) temperature range over which the glass transition takes place; the observed temperature can vary with the specific property chosen for observation and on details of the experimental procedure. The observed Tg should be considered only as an estimate. See also "heat distortion temperature".
GLYCIDYL ETHER GROUP	A characteristic functional group of epoxy resins, part of which is the epoxy group itself.

GROUT (RESIN)	A resin-based system, usually containing fillers, which may be pumped or poured into gaps or voids (a widely used term for other products, ie: cement grouts, chemical grouts, tile grouts).
GRP	Glass (fibre) reinforced plastic(s); may be prefixed by the name of the resin matrix, eg polyester GRP or phenolic GRP. (Associated term: FRP where fibres other than glass are used). In the polyester industry, it is taken to mean glass-reinforced polyester.
HARDENER	The cross-linking agent of a resin-based product. In two component products for example, the curing agent for an epoxy resin or the catalyst component of a polyester or similar resin; more generally, the component added to cause solidification of a reactive polymer.
HARMONISED EUROPEAN STANDARD	A European Standard which addresses the essential requirements of the CPD.
HEAT DISTORTION (DEFLECTION) TEMPERATURE	The temperature at which an arbitrary deformation occurs when a specimen (often a prism) is subjected to an arbitrary set of test conditions (often a flexural load). A number of standard test methods are available but different loads and rates of heating are applied and different limits of deformation are specified. Principally of value in control and development work although a reasonably linear relationship to the glass transition temperature may exist for some types of polymer.
HOMOPOLYMER	A polymer formed from one type of monomer.
HYDROLYSIS	Decomposition caused by the chemical action of water.
IMPREGNANT	A collective term for "penetrants" and "surface sealer".
INHIBITOR	A material used (usually) in low concentration to suppress a chemical reaction; usually applied to additives for polyester and similar resins which improve the shelf-life by delaying/preventing premature thermal or light-induced cross-linking.
INITIATOR	See "activator" and "catalyst".
INORGANIC POLYMER	A polymer which is not derived from compounds of carbon, eg silicates.
INTUMESCENT COATING	A coating that when heated by a flame expands into a relatively rigid foam and thus insulates and provides fire protection for the substrate.
LAMINATE	A product made by bonding together two or more layers of material(s) (lamina is a single ply or layer).
LATEX DISPERSION	An aqueous dispersion of natural or synthetic rubber.
MAIN CHAIN	The predominant structural part or backbone of a polymer.

MASTIC (CAULK)	A material which remains essentially paste-like beneath a surface skin which forms after application.
(METH)ACRYLATE POLYMERS	See "acrylate polymers".
MOISTURE CURING	A cross-linking reaction (intended to be) induced by atmospheric moisture.
MOLECULAR WEIGHT	A molecular mass, expressed on a relative scale, which has a unique value for most compounds; most linear polymers contain molecules having many different chain lengths and, therefore, they also have a distribution of molecular weights.
MONOMER	A relatively small molecule capable of reacting with like or unlike molecules to form a polymer.
MORTAR (RESIN)	A blend of fine sand ($\leq$5 mm) or other particulate filler, generally of graded composition, with a resin-based binder.
OLIGOMER	A substance composed of only a few monomeric units, eg a dimer, trimer; in contrast to polymers, the properties vary with the addition/removal of one monomeric unit; most commercial liquid epoxy resins are oligomers – chain extension, in addition to cross-linking, occurs during the cure reaction.
OPEN TIME	The maximum allowable time under specific conditions between mixing/applying an adhesive and closing a joint if reliable adhesion is to be obtained; also used where a primer must still be tacky when a second product, eg a mortar, is applied.
ORGANIC POLYMER	A polymer whose main chain is derived from compounds of carbon.
OSMOSIS	Liquid flow from a solvent to a solution separated by a solvent-permeable (semi-permeable) membrane. Many different materials (eg coatings) may act as such membranes.
PENDENT POSITION	The position of a chemical group which is attached to, but is not within, a main polymer chain (may also be referred to as a side-group).
PENETRANT	A material which penetrates and waterproofs porous materials by lining the pores (may also be referred to as an impregnant or surface sealer).
PIGMENT VOLUME CONCENTRATION (PVC)	The concentration, by volume, of a pigment (plus extender) in a paint/coating or in a dry film, usually expressed as a percentage of the non-volatile component (the critical PVC is the value at which the voids between the solid particles that are nominally touching are just filled with binder; in this region certain properties are changed markedly).
PLASTIC	A polymeric material, usually synthetic and containing other ingredients, which can be shaped by flow.

PLASTICISER	A general term for materials incorporated to increase flexibility in the cured state; also used, mostly for cementitious products, for materials which improve workability.
POLYESTERS	Linear thermoplastic polymers in which the repeat structural unit in the main chain is of the ester type, derived from reaction between an alcohol (a diol) and an organic (dicarboxylic) acid. The qualifying description unsaturated is commonly implied in the term "polyester resins" as the chains also contain C=C bonds which can act as functional groups for cross-linking; the presence of a cross-linking monomer/solvent, eg styrene, may also be assumed.
POLYMER	A substance with a relatively high molecular mass, consisting of one or more structural unit(s) repeated many times, these units being derived from monomers; (also known as high polymers and as macro or giant molecules). Organic polymers generally have main chains of various length which can be represented by an essentially linear molecular structure. These main chains may also have branches of varying length and complexity and/or they may be cross-linked.
POLYMER-MODIFIED CONCRETE (PMC)	A pre-mixed material in which either a non-reactive polymer (eg a latex dispersion) or a reactive monomer/resin is added to a fresh cement concrete (mortar) mix in a liquid, powder or dispersed form.
POLYMER (PORTLAND) CEMENT CONCRETE (P(P)CC)	A polymer modified concrete.
POLYMER (RESIN)-CONCRETE (PC)	A composite in which aggregate (filler) is bonded together by a polymer (resin).
POLYMER (RESIN)-IMPREGNATED CONCRETE (PIC)	A composite in which hardened concrete (mortar) is impregnated with a reactive monomer (resin) which is then polymerised in situ.
POLYURETHANES	Polymers formed by the reaction of an organic diisocyanate with compounds containing hydroxyl groups. Polyurethanes may be thermosetting, thermoplastic, elastomeric or cellular.
POST CURING	To increase the extent of cross-linking by heating above the initial cure temperature (commonly used to improve/optimise the properties of resin systems cured under factory conditions).
POT-LIFE	The period of time under specific conditions during which a reacting composition remains suitable for its intended purpose, ie after mixing with the reaction-initiating agents; also referred to as the useable life (the difference between pot-life and gel time should be noted – see "gel point").
PRE-ACCELERATED RESIN	A polyester or similar resin containing an accelerator incorporated by the manufacturer.

PRE-POLYMER	A polymer (or precursor) that is capable of further polymerisation by reaction with itself or with other polymers or monomers. Linear unsaturated polyesters are pre-polymers, cross-linked by a monomer (styrene) to give a larger molecular structure.
PRIMER	The first coat of a painting system that helps to bind subsequent coats to the substrate (and may inhibit its deterioration, eg by corrosion); also refers to resin products applied prior to a mortar, grout, concrete.
PROMOTER	See "accelerator".
RAW MATERIALS	The basic components in a formulation (the term does not imply a lack of processing).
REACTIVE POLYMER	A collective term for certain pre-polymers which can be further polymerised at normal pressures and temperatures to give a cross-linked structure.
RESIN (SYNTHETIC)	A heterogeneous group of chemically produced substances that resemble or share some of the physical characteristics of natural resins, eg viscosity for liquid resins (both types may also be solid or semi-solid); generally produced from simpler compounds by condensation reactions and/or polymerisation. In a broad sense, the term is used for any polymer that is a basic material for plastics. Commonly used to describe both a pre-polymer and the derived cross-linked polymer.
RETARDER	A substance added to slow down a chemical or physical change; solvents may retard chemical reactions simply by diluting the reactants.
RHEOLOGY	The study of the flow properties of materials.
ROVINGS	Strands of continuous fibres.
SEALANT	A material used to seal a designed joint and to accommodate movement while remaining bonded to the appropriate joint surfaces.
SELF-LEVELLING	A term applied to flooring compositions that, after being uniformly spread, flow to give a smooth surface without trowelling (more correctly self-smoothing).
SEMI-ORGANIC POLYMERS	Polymers having inorganic main chains but organic side-groups, eg silicones.
SHELF-LIFE	The time during which a material should remain in good condition if correctly stored; applies to both raw materials and formulated products and may be referred to specific properties such as reactivity, viscosity, colour.

SILANES	Compounds containing silicon and hydrogen; also (and more commonly) derivatives which can contain a wide variety of other atoms or groups in place of the hydrogen atom(s); also the monomers for silicone resins (the term organo-functional silane is used where the attached groups can be chemically linked to the polymeric matrix bonded to a surface).
SILICONES	A class of materials (which includes polymers) all of which contain alternating silicon and oxygen atoms in the main chain.
SILOXANES	May be regarded as (stable) intermediates in the manufacture or formation of silicone polymers from silanes; also the name given to the repeating silicon – oxygen group in a silicone (silicone polymers are in fact more correctly called polyorganosiloxanes).
SOLVENT	Any liquid in which another substance (a liquid, solid or gas) dissolves. Most organic solvents are volatile and possess a distinctive odour; many are flammable and toxic. In coatings, they are used to dissolve or disperse the constituents responsible for film formation and to control consistency.
STYRENE	The (unsaturated) monomer for polystyrene; also the most common cross-linking monomer for unsaturated polyester and vinyl ester resins (the styrene also acts as a solvent for the linear pre-polymer and gives these resins their characteristic odour).
SUBSTRATE	The material on which another is placed for the purpose of bonding or penetration.
SURFACE SEALER	A material which penetrates and waterproofs porous materials by filling the pores.
SURFACE TENSION	the force produced in the surface of a liquid due to unbalanced intermolecular forces; cohesive forces in a liquid cause it to behave as if it is contained in an elastic skin and thus to assume a spherical shape. An important property in adhesive bonding as these forces must be overcome by the liquid/solid interaction for spreading and wetting of a surface to take place.
SYNTACTIC	Materials consisting of hollow sphere fillers in a resin matrix (a type of cellular or foam material).
TACK COAT	A primer which is overcoated while still tacky.
TERPOLYMER	A polymer derived from three types of monomer, eg acrylonitrile -butadiene-styrene (ABS polymers).
THERMOPLASTIC	A polymer that can be reversibly softened by heating and hardened by cooling through a temperature range characteristic of the material; repeated reshaping by heat and pressure is possible as no chemical changes are involved in the process.

THERMOSETTING	Capable of being changed into a substantially infusible and insoluble material when cured by heat or chemical reaction (usually applied to pre-polymers that are cross-linked to give a thermoset). A thermosetting reaction may take place at ambient temperatures.
THIXOTROPE	An additive which specifically imparts "thixotropy".
THIXOTROPY	A property of a fluid material which is both reversible and time-dependent and which is normally associated with the formation of a specific internal structure. Thus, although flow takes place on application of a shear stress, a significantly more viscous, often gel-like, state is produced on standing; the time required for this transition – a characteristic of the material, may vary widely. Often used incorrectly for fluid/ plastic materials with a high viscosity or yield point. Useful in preventing drainage, sagging or dripping of a fluid.
UNSATURATED	A carbon chain in which there is at least one double (or triple) bond, represented as C=C; such bonds are uniquely reactive as the addition of other species can take place while a bond between the two original atoms is maintained.
USEABLE LIFE	See "pot-life".
VINYL ESTERS	A group of resin systems often based on a chemically complex main chain which may contain ester and urethane groups, some unsaturation, and groups derived from epoxies; the end-groups on the main chain of the linear pre-polymers are usually of the vinyl type (more specifically an acrylic ester (acrylate)) and act as functional groups for a cross-linking reaction.
VINYL GROUP	The functional group ($CH_2 = CH$-) characterising many different monomer types ($CH_2 = CH$ -X) used to form a very wide variety of polymers; a number of these polymers (eg polystyrene, acrylics and polyvinyl chloride) are commonly treated as distinct sub-groups.
VISCOELASTIC	A combination of viscous and elastic behaviour whereby response to an applied stress is time-dependent (or damped).
VISCOSITY	The internal resistance to flow or shear possessed by a liquid. For Newtonian liquids, the rate of flow (shear) is directly proportional to the shearing force and the viscosity, the ratio of these parameters, is thus a constant at a given temperature.

VULCANISATION	An irreversible process during which a rubber compound, through a change in its chemical structure (ie cross-linking), becomes less plastic and elastic properties are conferred, improved, or extended over a greater range of temperature; greater resistance to swelling by organic liquids is also achieved.
WOVEN ROVINGS	A fibre material (usually glass) made by the weaving of rovings.
ZINC-RICH PRIMER	An anti-corrosive coating material for iron and steel incorporating zinc dust in a concentration sufficient to give cathodic protection whereby the dry film is electrically conductive, enabling the zinc to corrode preferentially to the substrate.

References

ACTON, P (1979)
'Plane' answer to road repairs
Surveyor, 27 September

ALLEN, RTL, EDWARDS, SC and SHAW JDN (editors) (1993)
The repair of concrete structures (2nd edition)
Blackie Academic and Professional, Glasgow

AMERICAN CONCRETE INSTITUTE (1986)
Guide for the use of polymers in concrete (ACI Committee Report 548. IR-86; revised 1992)
ACI Manual of Concrete Practice

ANDREWS, EH (1968)
Fracture in Polymers
Oliver and Boyd, London

ANON (1980)
Space frame and Balvac repair subsided road
Construction News, 13 November

ANON (1982)
Remedial repairs on M40
Civil Engineering, August, 28

ANON (1985)
Woodweld winner at Dundee
New Civil Engineer, 11 July, 36

ANON (1994)
Trucks welcome on UK plastics span
Engineering News Record, 18 July, 22–23

ASTM A775/A775M-97 (1997)
Standard specification for epoxy coated reinforcing steel bars

BAILEY, M (1988)
The measurement of the slip resistance of floor surfaces – the Tortus and the pendulum
Construction and Building Materials, 2(3) 163–171

BARES, RA (1978)
Furane resin concrete and its application to large diameter sewer pipes
Polymers in Concrete (ACI Publication SP–58) 41–74
American Concrete Institute, Detroit

BARRETT, N (1985)
Resin fortifies listed timbers
New Civil Engineer, 9 May, 22–23

BASE, GD (1962)
Shear tests on very thin epoxy resin joints between pre-cast concrete units
Cement and Concrete Association Technical Report TRA/366, December

BEVERIDGE, RLW (1976)
Resin anchors
Civil Engineering, August

BOND, R (1980)
Concrete evidence of success
Surveyor, 11 September

BOWDEN, CJ (1993)
Restoration and repair of old timber
Construction Repair 7(6) November/December, 6–8

BOWDEN, CJ (1984)
Restoring life to old timbers
Building Trades Journal, 187 (5540), 2 February, 37

BOWDEN, CJ and SHAW, JDN (1987)
The use of polymers to repair ancient buildings
The Production, Performance and Potential of Polymers in Concrete,
ICPIC Brighton , 277–280

BOYES, R (1985)
Resin filled crack injection repairs
Civil Engineering, April, 24 and 27

BROMLEY, D and WOODWARD, JC (1973)
Use of epoxy resin formulations as self-levelling grouts
Symposium on Resins and Concrete, University of Newcastle-upon-Tyne, 17–18 April
The Plastics Institute/Institution of Civil Engineers

BS 476: 1970 onwards
Fire tests on building materials and structures (Parts 3–33)

BS 1243: 1978
Specification for metal ties for cavity wall construction

BS 2050: 1978 (1996)
Specification for electrical resistance of conducting and anti-static products made from flexible polymeric material

BS 2782: 1975 onwards
Methods of testing plastics (Parts 0–12)

BS 4247: 1981 and 1982
Surface materials for use in radioactive areas (Parts 1 and 2)

BS 5080: 1986 and 1993
Structural fixings in concrete and masonry (Parts 1 and 2)

BS 5350: 1976 onwards
Methods of test for adhesives (Part 0 and Groups a–h)

BS 5442: 1979 (1990) and 1989
Classification of adhesives for construction (Parts 1–3)

BS 5493: 1977
Code of practice for protective coating of iron and steel structures against corrosion

BS 5958: 1991
Code of practice for control of undesirable static electricity (Parts 1 and 2)

BS 6093: 1993
Code of practice for design of joints and jointing in building construction

BS 6319: 1983–1993
Testing of resin and polymer/cement compositions for use in construction (Parts 1–12)

BS 6920: 1996
Suitability of non-metallic products for use in contact with water intended for human consumption with regard to their effect on the quality of the water (Parts 1-3)

BS 7295: 1990
Resin-bonded epoxy coated carbon steel bars for the reinforcement of concrete (Parts 1 and 2)

BS 7543: 1992
Guide to the durability of buildings and building elements, products and components

BS 8081: 1989
Code of practice for ground anchorages

BS 8204: 1987–1994
Screeds, bases and in-situ floorings (Parts 1–5)

BS 8210: 1986
Guide to building maintenance management

BS EN 1081: 1998
Resilient floor-coverings: determination of electrical resistance

BUILDING RESEARCH ESTABLISHMENT (1982)
Installation of wall ties in existing construction
BRE Digest 257, CRC Ltd, London

BUILDING RESEARCH ESTABLISHMENT (1993)
Installing wall ties in existing construction
BRE Digest 329, CRC Ltd, London

BUILDING RESEARCH ESTABLISHMENT (1995)
Standardisation in support of European legislation: What does it mean for the UK construction industry?
BRE Digest 397, CRC Ltd, London

BUSSELL, MN and CATHER, R (1995)
Design and construction of joints in concrete structures, CIRIA Report 146
Construction Industry Research and Information Association, London

CHAPLIN, RG (1980)
Abrasion resistant concrete floors
Advances in Concrete Slab Technology, 532–543
Pergamon Press, Oxford

CHRISTIE, SH, McCLAIN, RR and MELLOAN, JH (1981)
Epoxy modified Portland cement concrete
Applications of Polymer Concrete (ACI Publication SP–69) 155–167
American Concrete Institute, Detroit

CHUNG, HW (1975)
Epoxy repaired reinforced concrete beams
ACI Journal, 72(5) May, 233–234

CHUNG, HW (1981)
Epoxy repair of bond in reinforced concrete members
ACI Journal, 78(7) Jan–Feb, 79–82

CIBA-GEIGY AG (1994)
World's longest composite bridge bonded with Araldite
CIBA Polymers – Aspects, Issue 34

CIRIA (1981)
Report of the civil engineering task force to the Science and Engineering Research Council and the Departments of Environment and Transport
Construction Industry Research and Information Association, London

CLARKE, JL (editor) (1996)
Structural design of polymer composites EUROCOMP design code and handbook.
E & FN Spon, London

COMITE EURO-INTERNATIONAL Du BETON (1994)
Fastenings to concrete and masonry structures – state of the art report
Thomas Telford Services Limited, London

DEPARTMENT OF TRANSPORT (1990)
Highways and traffic departmental standard BD43/90 – criteria and material for the impregnation of concrete highway structures
Department of Transport, London

DEPARTMENT OF TRANSPORT (1994a)
Highways and traffic departmental standard BD47/94: waterproofing and surfacing of concrete bridge decks
Department of Transport, London

DEPARTMENT OF TRANSPORT (1994b)
Strengthening of concrete highway structures using externally bonded plates
Advice Note BA30/94
Department of Transport, London

DE VEKEY, RC (1979a)
Corrosion of steel wall ties: recognition, assessment and appropriate action
BRE Information Paper 28/79, CRC Ltd, London

DE VEKEY, RC (1979b)
Replacement of cavity wall ties using resin grouted stainless steel rods
BRE Information Paper 29/79, CRC Ltd, London

DIKEOU, JT (1987)
Pre-cast polymer concrete in the United States
The Production, Performance and Potential of Polymers in Concrete, ICPIC, Brighton 251–256
Brighton Polytechnic, Sussex, England

DIKEOU, JT and KAEDING, AO (1992)
U.S. and other specifications and standards for polymer concretes
Polymers in Concrete, ICPIC, Moscow, 9–25
Science Business Contact International

DUMBLETON, B (1989)
Stuck fast
New Civil Engineer, 16 March, 36–37

DUNSTER, JA and HOUGH, NE (1968)
Constructing Rawcliffe Bridge
Civil Engineering and Public Works Review, June, 651–655

DURST, WJ (1974)
The Shellgrip surfacing system
Highways and Road Construction, May 12–16

DUSSEK, IJ (1980)
Strengthening of bridge beams and similar structures by means of epoxy resin bonded external reinforcement
TRB Record 785, 21–24, Transportation Research Board, USA

EC DIRECTIVE: EEC/93/43 (19 July 1993)
Hygiene of foodstuff
HMSO

ELLIOTT, IH and SPOLTON, RL (1985)
Greater Cairo wastewater project: design of east bank stage I, works tunnels
Proc. Instn Civil Engineers, Part I, 76, 807–829

EMBERSON, NK and MAYS, GC (1987)
Polymers mortars for the repair of structural concrete: the significance of property mismatch
The Production, Performance and Potential of Polymers in Concrete,
ICPIC Brighton, 335–341
Brighton Polytechnic, Sussex, England

EMBERSON, NK and MAYS, GC (1990a)
Design of patch repairs: measurements of physical and mechanical properties of repair systems for satisfactory structural performance
International Conference, Protection of Concrete, University of Dundee,
11–13 September, 937–954
E & FN Spon

EMBERSON, NK and MAYS, GC (1990b)
Significance of property mismatch in the patch repair of structural concrete. Part 1: properties of repair systems. Part 2: axially loaded reinforced concrete members
Magazine of Concrete Res. 42(152), 147–170

EUROPEAN UNION OF AGRÉMENT (1992)
(i) MOAT No. 49, UEAtc technical guide on anchors for use in cracked and non-cracked concrete
(ii) MOAT No. 42, UEAtc directives for assessment of anchor bolts
British Board of Agrément, Watford, UK

FARMER, N (1994)
Plate bonding – (1) design aspects
Construction Repair, 8(1), Jan/Feb, 41–42

FeRFA (1990)
Osmosis in flooring (technical report)
FeRFA, Aldershot, UK

FeRFA (1991a)
Resin mortar bedding for bridge bearing pads: application guide no. 1
FeRFA, Aldershot, UK

FeRFA (1991b)
Synthetic resin flooring: application guide no. 4
FeRFA, Aldershot, UK

FIP (1978)
Proposal for a standard for acceptance tests and verification of epoxy bonding agents for segmented construction: FIP/9/2
Fédération Internationale de la Précontrainte/Cement and Concrete Research Association, Crowthorne

FIP COMMISSION ON PRACTICAL CONSTRUCTION (1991)
Repair and strengthening of concrete structures – FIP guide to good practice
Thomas Telford, London

FLEMING, CJ and KING, G (1967)
The development of structural adhesives
RILEM International Symposium: Experimental Research on New Developments Brought by Synthetic Resins to Building Techniques, Paris, September 1967, 241–251

FONTANA, JJ (1987)
Electrically conductive polymer concrete coatings
Polymer Modified Concrete (ACI Publication SP–99), 31–50
American Concrete Institute, Detroit

FONTANA, JJ and MENDIS, P (1989)
Installation of an insulating polymer concrete overlay for dike insulation
Polymers in Concrete: Advances and Applications (ACI Publication SP-116) 145–156
American Concrete Institute, Detroit

FONTANA, JJ, REAMS, W and CHENG, HC (1987)
Potential polymer concrete heat exchanger tubes for corrosive environments
The Production, Performance and Potential of Polymers in Concrete,
ICPIC, Brighton, 399–403
Brighton Polytechnic, Sussex, England

FONTANA, JJ, REAMS, W and ELLING, D (1989)
Electrically conductive polymer concrete overlay installed in Palaski, Virginia
Polymers in Concrete: Advances and Applications (ACI Publication SP-116), 157–176,
American Concrete Institute, Detroit

FONTANA, JJ, STEINBERG, M and REAMS, W (1987)
Lightweight polymer concrete composites
Polymer Modified Concrete (ACI Publication SP-99) 51–72
American Concrete Institute, Detroit

FOWLER, D and HSU, H T (1984)
Static and cyclic behaviour of polymer concrete beams
Polymers in Concrete ICPIC, Darmstadt, Germany, 159–164
Institut für Spanende Technologie und Werkzeugmaschinen, Darmstadt

FOWLER, DW (1987)
Current status of polymers in concrete
The Production, Performance and Potential of Polymers in Concrete, ICPIC, Brighton, 3–7,
Brighton Polytechnic, Sussex, England

FOWLER, DW (1989)
Future trends in polymer concrete
Polymers in Concrete: Advances and Applications (ACI Publication SP-116), 129–143,
American Concrete Institute, Detroit

FOWLER, DW (1991)
Structural design of polymer concrete
Polymers in Concrete, ICPIC, North American Workshop, San Francisco, September

FRIDLEY, KJ, WHEAT, DL and FOWLER, DW (1989)
Sandwich beams with polymer concrete facings
Polymers in Concrete: Advances and Applications (ACI Publication SP-116), 35–59,
American Concrete Institute, Detroit

GALKA, R and YATES, APJ (1984)
Pipe protection – a review of current practice
The British Hydromechanics Research Association (with the Paint Research Association) Bedford

GATFIELD, MJ (1997)
Screeds and floorings. A guide to their selection, construction, testing and maintenance
CIRIA Report R184
Construction Industry Research and Information Association, London

GAUL, RW (1990)
Application of bonded plate reinforcement in North America
International Seminar on Structural Repairs/Strengthening by the Plate Bonding Technique
University of Sheffield, September

GILES, CG, SABEY, BE and CARDEW, KMF (1964)
Development and performance of the portable skid resistance tester
Road Research Technical Paper No. 66, HMSO

GILMOUR, DK (1969)
Application of resin surfacing to steel bridge decks
Epikote in the Construction Industries, Symposium organised by Shell Chemicals, UK

GOSNEY, J (1974)
Rock steady at Edinburgh Castle
Contract Journal, 22 August 1974, 28

GREATER LONDON COUNCIL (1971)
Slip reistance of floors, stairs and pavings
Bulletin No. 43 (2nd series), Item No. 5, March

GUTHRIE, JM and MILNE, J (1983)
Vacuum grouting of concrete road slabs
Journal of the Institution of Highway Engineers, May

HALL, AF (1973)
The use of resin-based grouts for machine mounting and installation of structures
Symposium on Resins and Concrete, University of Newcastle upon Tyne, 17–18 April, 1973
The Plastics Institute/Institution of Civil Engineers

HARBRON, R (1994)
Advances in shrinkage control provide new techniques in rapid floor reinstatement
Construction Repair, 8(2) 38–40

HATHERLY, LW, MAHAFFY, JH and TWEDDLE, A (1969)
The skid resistance of city streets and road safety
Journal of the Institution of Highway Engineers, April, 7–14

HAYWARD, D (1982)
Programme shift at Windscale ponds
New Civil Engineer, 24 June

HEAD, PR (1994)
Cost-effective advanced composite structures designed for life in the infrastructure (Paper 13)
19th International BPF Composites Congress, Birmingham, November 1994, BPF Publication No. 293/8, 59–73

HEAYES, N (1979)
Balvac speeds up repair jobs
Contract Journal, 8 March

HEWLETT, PC and HURLEY, SA (1985a)
Repair materials: selection and testing
Second International Conference on Structural Faults and Repair, London,
30 April–2 May, 219–232
Engineering Technics Press Ltd

HEWLETT, PC and HURLEY, SA (1985b)
The consequences of polymer-concrete mismatch
Design life of buildings – an ICE symposium, November 1984, 179–196
Thomas Telford, London

HEWLETT, PC and MORGAN, JG (1982)
Static and cyclic response of reinforced concrete beams repaired by resin injection
Magazine of Concrete Research, 34(118), 5–17

HEWLETT, PC and SHAW, JDN (1977)
Structural adhesives in civil engineering
Developments in Adhesives (WC Wake, editor), Chapter 2, 25–75
Applied Science Publishers, New York

HEYWOOD, D (1974)
Epoxy resin mortar jointings of concrete piles
DoE Construction, 11, September, 12–14

HIGHWAYS AGENCY (1991 and 1993 amendments)
Specification for highway works: 900 series: Surfaces (Clause 924)
HMSO

HUGENSCHMIDT, F (1975)
Epoxy adhesives for concrete and steel
Polymers in Concrete, ICPIC, London, May 1975, 195–209
The Construction Press (for the Concrete Society)

HUGENSCHMIDT, F (1982)
New experiences with epoxies for structural applications
International Journal of Adhesion and Adhesives, 2(2), April, 84–96

HUTCHINSON, AR (1995)
Forensic studies of adhesive joints, Part 6 – externally bonded steel plate reinforcement for concrete bridges
The performance of adhesive joints – research programme funded by the Department of Trade and Industry, Project 3, Report 9

HUTCHINSON, AR and RAHIMI, H (1993)
Behaviour of reinforced concrete beams with externally bonded fibre-reinforced plastics
5th International Conference on Structural Faults and Repair, University of Edinburgh, July, 221–228 Engineering Technics Press

IINO, T and OTOKAWA, K (1981)
Application of epoxy resins in strengthening of concrete structures
Polymers in Concrete, ICPIC, Koriyama, Japan, 997–1011
Nihon University, Japan

IMAMURA, KI, NAKANO, M and TSUJI, K (1981)
Resin concrete products in NTT
Polymers in Concrete, ICPIC, Koriyama, Japan, 88–98
Nihon University Japan

IMAMURA, K, TOYOKAWA, K and MURAI, N (1978)
Pre-cast polymer concrete for utilities applications
Polymers in Concrete, ICPIC, Austin, Texas, 173–186
The University of Austin, USA

JAMES, I (1987)
Measurement of slip-resistance: a modern approach to standardisation
Chemistry and Industry, 2 February, 89–94

KAEDING, AO (1985)
Corrosion resistant pipe liners – polymer concrete
Polymer Concrete – Uses, Materials and Properties (ACI Publication SP-89) 19–42
American Concrete Institute, Detroit

KAEDING, AO (1991)
Design criteria for pre-cast polymer concrete
Polymers in Concrete, ICPIC North American Workshop, San Francisco, September

KANE, JF (1984)
New applications for polymer concrete in bridge and tunnel rehabilitation
Polymers in Concrete, ICPIC, Darmstadt, Germany, 447–450
Institut für Spanende Technologie und Werkzeugmaschinen, Darmstadt

KAY, EA, STAYNES, BW, MAYS, GC, TILLYER, B, WOOD, JGM, KING, EC and DUNNE, HB (1993)
Standard tests for repair materials and coatings for concrete
CIRIA Technical Note 141
Construction Industry Research and Information Association, London

KEER, JG and EMBERSON, NK (1986)
A theoretical and experimental study of concrete members repaired by resin injection
Adhesion between polymers and concrete (ISAP 86), RILEM international symposium, Aix-en-Provence, France, September 1986, 419–431
Chapman and Hall, London

KETTLE, R and SADEGZADEH, M (1986)
Indirect and non-destructive methods for assessing abrasion resistance of concrete
Mag. Conc. Res. 38(137) 183–190

KETTLE, RJ and SADEGZADEH, M (1986)
Recent research developments in abrasion resistance
Concrete, November, 29–31

KETTLE, RJ and SADEGZADEH, M (1987)
Abrasion resistance of polymer impregnated concrete
Concrete, May, 32–34

KLAIBER, FW, DUNKER, KF, WIPF, TJ and SANDERS, WW (1987)
Methods of strengthening existing highway bridges
National Cooperative Highway Research Program Report 293
Transportation Research Board, Washington, DC

KOBAYASHI, H, ONO, H and HIROSE, T (1981)
Quick setting mortar lining process
Polymers in Concrete, ICPIC, Koriyama, Japan, 584–598
Nihon University, Japan

KOYANAGI, W, ROKUGO, K, UCHIDA, Y et al (1990)
Structural behaviour of unsaturated polyester resin concrete slabs
Polymers in Concrete, ICPIC, Shanghai, 405–410
International Academic Publishers

KREIS, R (1981)
Continuing manufacturing process for polymer concrete and newest applications
Polymers in Concrete, ICPIC, Koriyama, Japan, 676–689
Nihon University, Japan

KREIS, R (1987)
Preparation methods for polymer concrete and mortar
The Production, Performance and Potential of Polymers in Concrete, ICPIC, Brighton, 133–137
Brighton Polytechnic, Sussex, England

KREIS, R (1990)
Production facilities of polymer concrete and mortar
Polymers in Concrete, ICPIC, Shanghai, 773–779
International Academic Publishers

KREIS, R (1992)
Mass production of polymer concrete
Polymers in Concrete, ICPIC, Moscow, 278–289
Science Business Contact International

KRIEGH, JD, NORDBY, GM and ENDEBROCK, EG (1967)
Use of epoxy compounds as a shear connector in composite T-beams
Highway Research Record No. 155, 1–13

LALIBERTE, J (1984)
Application of polymer concrete in Canada
Polymers in Concrete, ICPIC, Darmstadt, Germany, 45–52
Institut für Spanende Technologie und Werkzeugmaschinen, Darmstadt

LEEMING, MB (1996)
A ROBUST solution to strengthening RC and PC beams
Construction Repair, 9(5) 15–17

LEEMING, MB and PESHKAM, V (1995)
A ROBUST solution to strengthening bridges
6th International Conference on Structural Faults and Repair, Westminster, London, July, 161–164, Engineering Technics Press

LEWIS, AF and RAMSEY, WB (1966)
Mechanical behaviour of polymers and adhesive joint strengths with amine cured epoxy resins
Adhesives Age, February, 9(2), 20–27

L'HERMITE, R and BRESSON, J (1967)
Concrete strengthened by glued reinforcement
RILEM International Symposium: Experimental Research on New Developments Brought by Synthetic Resins to Building Techniques, Paris, September, 1967, 175–203

LIN, B, LU, A and CENG, R (1986)
Study of 3200 vinyl ester resin mortar and its applications.
Adhesion between Polymers and Concrete: Proceedings of an International Symposium organised by RILEM Technical Committee 52, Aix-en-Provence, 16–19 September 1986
Sasse, H R (editor) 245–254
Chapman and Hall, London/New York

MACDONALD, MD (1974)
Waterproofing concrete bridge decks: materials and methods
TRRL Laboratory Report 636
Transport and Road Research Laboratory, Crowthorne, Berks

MAEDA, Y, MATSUI, S, SHIMADA, I and KATO, H (1980)
Deterioration and repairing of reinforced concrete slabs of highway bridges in Japan
Technology Reports (Osaka University) 30 (1599) 135–144

MAKANSI, I and READER, CEL (1975)
Epoxy water slurry mortars
Polymers in Concrete, ICPIC, London, 5–7 May, 174–178
The Construction Press (for the Concrete Society)

MANDER, RF (1981)
Use of resins in road and bridge construction and repair
International Journal of Cement Composites and Lightweight Concrete, 3(1), 27–39

MANNING, DG (1996)
Corrosion performance of epoxy-coated reinforcing steel: North American experience
Construction and Building Materials 10(5) 349–365

MARTIN, DM (1992)
Tests on bonded transverse intermediate web stiffeners
The Structural Engineer, 70(15), 261–267

MATHEWS, R (1983)
Hidden liability
Building, 4 November 1983, 52–53

MATSUURA, T and KONAKA, T (1981)
Quick setting resin mortar
Polymers in Concrete, ICPIC, Koriyama, Japan, 569–583
Nihon University, Japan

MAY, CA and TANAKA, Y (1973)
Epoxy resins – chemistry and technology
Dekker, New York

MAYS, GC (1985)
Structural applications of adhesives in civil engineering
Materials Science and Technology, Vol 1, 937–943

MAYS, GC (editor) (1992)
Durability of concrete structures
E & FN Spon, London

MAYS, GC and HUTCHINSON, AR (1988)
Engineering property requirements for structural adhesives
Proc. Instn. Civ. Engrs., Part 2, 85, 485–501

MAYS, GC and HUTCHINSON, AR (1992)
Adhesives in Civil Engineering
Cambridge University Press, Cambridge

MAYS, GC and VARDY, AE (1982)
Adhesive-bonded steel/concrete composite construction
International Journal of Adhesion and Adhesives, 2(2), 103–107

MCCONNELL, WR (1966)
Epoxy surface treatments for Portland cement concrete pavements
Epoxies with Concrete (ACI Publication SP-21), 9–18
American Concrete Institute, Detroit

MCKENNA, JK and ERKI, MA (1994)
Strengthening of reinforced concrete flexural members using externally applied steel plates and fibre composite sheets – a survey
Can. J. Civ. Eng. 21, 16–24

MEIER, U (1992)
Carbon fibre reinforced polymers: modern materials in bridge engineering
Structural Engineering International, Vol 2, 7–12

MENDIS, P (1987)
Polymer concrete overlays and applications
The Production Performance and Potential of Polymers in Concrete, ICPIC, Brighton, 53–57
Brighton Polytechnic, Sussex, England

METTEM, CJ and DAVIS, G (1996a)
Resin-bonded repair systems for structural timber, Part 1
Construction Repair, 11(2), 23–28

METTEM, CJ and DAVIS, G (1996b)
Resin-bonded repair systems for structural timber, Part 2
Construction Repair, 11(4), 43–47

MIKHAILOV, KV, PATUROEV, VV and KREIS, R (1992)
Polymer concretes and their structural uses
AA Balkenna Publishers, Rotterdam

MINISTRY OF TRANSPORT (1965)
Technical memorandum (Bridges) BEI: Waterproofing of bridge decks
Ministry of Transport, London

MINISTRY OF TRANSPORT (1970)
Technical Memorandum (Bridges) No. BE27
Waterproofing and surfacing of bridge decks
Ministry of Transport, London

MOORE, JFA (1981)
The performance of cavity wall ties
BRE Information Paper 4/81, CRC Ltd, London

NAVARATNARAJAH, V (1983)
Splicing of reinforcement bars with epoxy joints
International Journal of Adhesion and Adhesives 3(2) 93–99

NOJIRI, Y, KOBAYASHI, H, TAKATSUKA, T and HIROSE, T (1981)
The strength of the quick-setting resin mortar tunnel
Polymers in Concrete, ICPIC, Koriyama, Japan, 599–613
Nihon University, Japan

NUTT, WO (1967)
Inorganic polymer structures
British Patent No. 1 065 053 (April 1967)

NUTT, WO (1968)
Polymer cements – a review of some applications to concrete pavements
Paper given to Society of Chemical Industry, London (1968) and Cardiff (1969)

NUTT, WO (1969)
Stable composition of hydraulic cement and polymerisable unsaturated polyesterresin components copolymerisable by the addition of water and methods of producing same
US Patent 3,437,619 (8 April 1969)

NUTT, WO (1985)
Polymer concretes – a status review
Concrete, March, 33–34

NUTT, WO (1987)
State of the art in Europe
The Production, Performance and Potential of Polymers in Concrete, ICPIC, Brighton, 9–11
Brighton Polytechnic, Sussex, England

NUTTALL, N (1994)
Engineers test plastic road bridge
The Times, 30 June, 4

O'BRIEN, T (1973)
Jointing structural pre-cast concrete with resin adhesives: a review of progress
Symposium on Resins and Concrete, University of Newcastle upon Tyne, 17–18 April
The Plastics Institute/Institution of Civil Engineers

O'BRIEN, T (1984)
Resins in construction: feedback from structural applications
FeRFA Seminar: Resins in construction – 20+ years experience, London, 25 October

OHAMA, Y (1978)
Development of concrete-polymer materials in Japan
Polymers in Concrete, ICPIC, Austin, Texas, 121–137
The University of Austin, USA

OHAMA, Y (1991)
Worldwide industry activities, standardisation work, and recent trends in Japan with concrete-polymer composites
Polymers in Concrete, ICPIC, North American Workshop, San Francisco, September

OKADA, K, MIYAGAWA, T and KOYANAGI, W (1987)
Design recommendations for polyester resin concrete structures
The Production, Performance and Potential of Polymers in Concrete, ICPIC, Brighton, 81–83
Brighton Polytechnic, Sussex, England

OKADA, K and OHAMA, Y (1981)
Status of concrete-polymer composites in Japan
Polymers in Concrete, ICPIC, Koriyama, Japan, 3–19
Nihon University, Japan

OLIVER, A (1994)
Steel buildings to be glued together
New Builder, No. 236, 11 November, 3

PATERSON, R (1994)
Wall tie failure – developing an effective solution
Construction Repair 8(2) 48–50

PATERSON, WS (1976)
Loading tests on fixings in concrete
CIRIA Technical Note No 75
Construction Industry Research and Information Association, London

PATERSON, WS (1978)
Indicative fire tests on fixings, CIRIA Technical Note No. 92
Construction Industry Research and Information Association, London

PESHKAM, V and LEEMING, MB (1994)
Application of composites to strengthening of bridges: Project ROBUST (Paper 30)
19th International BPF Composites Congress, Birmingham, November, 1994, BPF Publication No 293/8

PHILLIPS, MW and SELWYN, JE (1978)
Epoxies for wood repairs in historic buildings
US Dept. of the Interior, Washington, DC

PLASTICS AND RUBBER WEEKLY (1993)
PRW Awards 1993 Supplement

PLUM, DR (1990)
The behaviour of polymer materials in concrete repair, and factors influencing selection
The Structural Engineer, 68(17), 337–345

POMEROY, CD (1992)
Concrete, an overview
Construction Materials Reference Book, Chapter 14, Doran, D K (editor)
Butterworth–Heinemann Ltd, Oxford

POPOVICS, S (1985)
Modification of Portland cement concrete with epoxy as admixture
Polymer Concrete – Uses, Materials and Properties (ACI Publication SP-89) 207–229
American Concrete Institute, Detroit

POTTER, WG (1975)
Uses of epoxy resins
Newnes-Butterworths, London

PRICE, AR (1990a)
A field trial of waterproofing systems for concrete bridge decks
Protection of Concrete, International Conference, University of Dundee, September 1990, Proceedings 367–380
E & FN Spon

PRICE, AR (1990b)
Laboratory tests on waterproofing systems for concrete bridge decks: TRRL Research Report 248
Transport and Road Research Laboratory, Crowthorne

PRUSINSKI, RC (1978)
Study of commercial development in pre-cast polymer concrete
Polymers in Concrete (ACI Publication SP-58) 75–101
American Concrete Institute, Detroit

PRUSINSKI, RC (1991)
Current status of pre-casting polymer concrete products in the building and transportation industry
Polymers in Concrete, ICPIC, North American Workshop, San Francisco, 24–27 September

QUINN, J and GODFREY, P (1991)
Development of a design code for structural polymer composites
Construction and Building Materials, 5(4), 193–200

RAITHBY, KD (1980)
External strengthening of concrete bridges with bonded steel plates
Report SR612, Transport and Road Research Laboratory, Crowthorne, Berks

RAYMOND, J (1990)
Overview of polymer concrete in the European Community
Paper presented to 6th ICPIC, September, Shanghai

REINER, M (1990)
Flexible rail fixing on bridges
Sika Information, October, 41–43

REINER, M (1991)
Elastic rail fixing in tunnels
Sika Information, October, 68–69

RESEARCH STRATEGY COMMITTEE FOR THE BUILDING AND CIVIL ENGINEERING ECONOMIC DEVELOPMENT COMMITTEE (1985)
Strategy for Construction R&D
National Economic Development Office, London

RICHARDS, R (1994)
Anti-static and conductive floors
Construction Repair 8(2), 30–31

RICHMOND, B (1993)
Bridge construction and beyond
Reinforced Plastics, October, 26–30

RIDOUT, G (1982)
Roads repaired at half price by Balvac
Contract Journal, 13 May, 14–15

ROBBINS, J (1991)
Premium bond
New Civil Engineer, 17 October, 18–19

ROBERTS, DG and SMITH, JG (1994)
High performance flooring
Construction Repair, July/August, 23–25

ROPER, P (1984)
Epoxy resins used to restore fine medieval timbers
Building Trades Journal, 29 March, 14–15

ROSENBERG, AM and CHAPMAN, DA (1991)
Improved unsaturated polyester resin concrete with shrinkage control
Polymers in Concrete, ICPIC, North American Workshop, San Francisco, 24–27 September

SAKAMOTO, O (1984)
A new continuous spraying system for resin mortar
Polymers in Concrete, ICPIC, Darmstadt, Germany, 293–298
Institut für Spanende Technologie und Werkzeugmaschinen, Darmstadt

SASSE, HR and HONSINGER, D (1990)
Department and optimisation of impregnation materials
Protection of Concrete, International Conference, University of Dundee, September 1990, Proceedings 177–190
E & FN Spon

SAWAIDE, M and TAKANO, I (1981)
Application of foamed epoxy resin in the construction field
Polymers in Concrete, ICPIC, Koriyama, Japan, 1028–1042
Nihon University, Japan

SCANLON, JM (1981)
Applications of concrete polymer materials in hydrotechnical construction
Applications of Polymer Concrete (ACI Publication SP-69) 45–62
American Concrete Institute, Detroit

SCARPINATO, ES (1984)
Epoxy bridge deck overlays – the Brooklyn bridge
Polymers in Concrete, ICPIC, Darmstadt, Germany, 351–358
Institut für Spanende Technologie und Werkzeugmaschinen, Darmstadt

SCHORN, H (1985)
Epoxy modified shotcrete
Polymer Concrete – Uses, Materials and Properties (ACI Publication SP-89) 249–260
American Concrete Institute, Detroit

SCHUTZ, RJ (1976)
Epoxy adhesives in prestressed and pre-cast concrete bridge construction
ACI J., 73(14) March, 155–159

SCOSS (1994)
Tenth report of the Standing Committee on Structural Safety
July 1992–June 1994, London

SHARP, WH (1981)
Restoration, upgrading and strengthening of ancient and historical timbers
Public Works Congress (Institution of Structural Engineers), 24th February

SHAW, JDN (1982)
A review of resins used in construction
International Journal of Adhesion and Adhesives, 2(2), 77–83

SHAW, JDN (1992)
Adhesives
Construction Materials Reference Book (editor D K Doran) Chapter 8,
Butterworth-Heinemann Ltd, Oxford

SHAW, JDN (1993)
Polymers for concrete repair
The Repair of Concrete Structures, 2nd edn (editor RTL Allen, SC Edwards and J D N Shaw) Chapter 4
Blackie Academic and Professional, Glasgow

SHAW, JDN (1994)
Review of industrial flooring
Construction Repair, 8(2), March/April, 32–37

SIMPSON, J (1990)
Hidden dangers
Building Today 22 March, 20–22

SIMS, FA (1979)
The use of resins in bridges and structural engineering in West Yorkshire
FeRFA Seminar: The use of resin in concrete construction and repair, London, 18 October 1979

SIMS, FA and WOODHEAD, S (1968)
Rawcliffe Bridge in Yorkshire
Civil Engineering and Public Works Review, April, 385–391

SPRINKEL, MM (1989)
Performance of multiple layer polymer concrete overlays on bridge decks
Polymers in Concrete: Advances and Applications (ACI Publication SP-116), 61–95
American Concrete Institute, Detroit

SRISKANDAN, K (1989)
Prestressed concrete road bridges in Great Britain: a historical survey
Proc. Instn. Civ. Engrs. Part 1, 86, 269–303

STAYNES, BW (1981)
Vacuum-impregnated epoxide resin concrete
Applications of Polymer Concrete (ACI Publication SP-69), 189–204
American Concrete Institute, Detroit

STAYNES, BW and ROSS, JH (1980)
Epoxide resin concrete – a practical application in superconducting turbogenerators
The Structural Engineer, 58A(7), 213–220

STUMES, P (1979)
W E R (wood-epoxy-reinforcement) – system manual; structural rehabilitation of deteriorated timber
Association for Preservation Technology, Dept. of Indian Affairs and Northern Development, Canada

TABOR, L J (1978)
Effective use of epoxy and polyester resins in civil engineering
CIRIA Report 69
Construction Industry Research and Information Association, London

TAKATSUKA, T, OKADA, T, MATSUURA, T and NAKANISHI, S (1984)
Quick-setting resin mortar for small diameter tunnel lining
Polymer in Concrete, ICPIC, Darnstadt, Germany, 195–200
Institut für Spanende Technologie und Werkzeugmaschinen, Darmstadt

TAPPIN, RGR, DOWLING, PJ and CLARK, PJ (1984)
Design and model testing of the Thames Gates
The Structural Engineer 62A(4), 115–124

TAYLOR, HPJ and RAYNER, MS (1989)
The Bomac System – history and recent developments
Paper presented to the Sheffield section of the Permanent Way Institution, 10 January

THE CONCRETE SOCIETY (1992)
Non-structural cracks in concrete (Technical Report no. 22 – 3rd edition)
The Concrete Society, Crowthorne

THE CONCRETE SOCIETY (1994)
Concrete industrial ground floors (Technical Report no. 34 – 2nd edition)
The Concrete Society, Crowthorne

TILLYER, RB (1990)
Upgrading existing floor surfaces
Concrete Society Seminar: Industrial Concrete Ground Floors
Update, 24 September

TUTT, JN (1988)
A guide to tie spacing and selection
CIRIA Report 117
Construction Industry Research and Information Association, London

VAN GEMERT, D (1986)
Long-term performance of epoxy bonded steel-concrete joints
Adhesion between polymers and concrete (ISAP86),
RILEM international symposium, Aix-en-Provence, France, September, 518–527
Chapman and Hall, London

VAN GEMERT, D (1990)
Evolutions in design and application of the plate bonding technique
Int. Seminar on Structural Repairs/Strengthening by the Plate Bonding Technique
University of Sheffield, September 1990

WOOD, JGM (1990)
Concrete repair materials for effective structural applications
Construction and Building Materials, 4(2), 64–67

WOODWARD, RJ and WILLIAMS, FW (1988)
Collapse of Ynys-y-Gwas bridge, West Glamorgan
Proc. Instn. Civ. Engrs. Part 1, 84, August, 635–669

YAMAMOTO, T (1987)
Corrosion control by pre-cast polymer concrete and carbon fiber reinforced coal ash sand for resin lining
The Production, Performance and Potential of Polymers in Concrete,
ICPIC, Brighton, 395–398
Brighton Polytechnic, Sussex, England

YAPING, X, HEYUN, C and RUILIANG, F (1987)
The study of PC-GRC composite for decoration
The Production, Performance and Potential of Polymers in Concrete,
ICPIC, Brighton, 257–263
Brighton Polytechnic, Sussex, England

YEON, K-S, FOWLER, DW and WHEAT, DL (1987)
Static flexural behaviour of various polymer concrete beams
The Production, Performance and Potential of Polymers in Concrete,
ICPIC, Brighton, 85–89
Brighton Polytechnic, Sussex, England

YIUN-YUAN, H (1990)
Polymerising the concrete
Polymers in Concrete, Proceedings of 6th ICPIC, Shanghai
International Academic Publishers

ZANKE, U (1989)
Computer-aided design of pre-cast polymer concrete line-drainage channels
Polymers in Concrete: Advances and Applications (ACI Publication SP-116) 15–26
American Concrete Institute, Detroit

ZELDIN, AN, KUKACKA, LE and CARCIELLO, N (1981)
New, novel well-cementing polymer concrete composite
Applications of Polymer Concrete (ACI Publication SP-69) 73–92
American Concrete Institute, Detroit

Bibliography

Additional bibliographies are provided in *Volume 2: Specification and use of the materials* (CIRIA PR78) and *Volume 3: Materials technology* (CIRIA PR79)

ALLEN, RTL, EDWARDS, SC and SHAW, JDN (editors) (1993)
The repair of concrete structures (2nd edition)
Blackie Academic and Professional, Glasgow

AMERICAN CONCRETE INSTITUTE (1968)
Expoxies with concrete (ACI Publication SP-21)
American Concrete Institute, Detroit

AMERICAN CONCRETE INSTITUTE (1973)
Polymers in concrete (ACI Publication SP-40)
American Concrete Institute, Detroit

AMERICAN CONCRETE INSTITUTE (1978)
Polymers in concrete (ACI Publication SP-58)
American Concrete Institute, Detroit

AMERICAN CONCRETE INSTITUTE (1981)
Applications of polymer concrete (ACI Publication SP-69)
American Concrete Institute, Detroit

AMERICAN CONCRETE INSTITUTE (1985a)
A guide for the use of waterproofing, dampproofing, protective and decorative barrier systems for concrete ACI 515.IR-79 (revised 1985)
ACI Manual of Concrete Practice

AMERICAN CONCRETE INSTITUTE (1985b)
Polymer concrete – uses, materials and properties (ACI Publication SP-89)
American Concrete Institute, Detroit

AMERICAN CONCRETE INSTITUTE (1987)
Polymer modified concrete (ACI Publication SP-99)
American Concrete Institute, Detroit

AMERICAN CONCRETE INSTITUTE (1989)
Polymers in concrete: advances and applications (ACI Publication SP-116)
American Concrete Institute, Detroit

AMERICAN CONCRETE INSTITUTE (1992a)
Guide for the selection of polymer adhesives with concrete ACI 503.5R-92
ACI Manual of Concrete Practice
American Concrete Institute, Detroit

AMERICAN CONCRETE INSTITUTE (1992b)
Guide for the use of polymers in concrete
ACI 548.IR-92
ACI Manual of Concrete Practice

AMERICAN CONCRETE INSTITUTE (1993a)
Causes, evaluation and repair of cracks in concrete structures
ACI 224.IR-93
ACI Manual of Concrete Practice

AMERICAN CONCRETE INSTITUTE (1993b)
Use of epoxy compounds with concrete
ACI 503R-93
ACI Manual of Concrete Practice

AMERICAN CONCRETE INSTITUTE (1996)
Properties and uses of polymers in concrete
ACI Publication SP-166
American Concrete Institute, Detroit

BAYLISS, DA and CHANDLER, KA (1991)
Steelwork corrosion control
Elsevier Science Publishers Ltd, Essex

BUILDING RESEARCH ESTABLISHMENT (1979)
Cellular plastics for building, Digest 224
CRC Ltd, London

CALDER, AJJ (1988)
Exposure tests on externally reinforced concrete beams performance after 10 years
Transport and Road Research Laboratory, Report 129

CHANDLER, KA (1985)
Marine and offshore corrosion
Butterworth & Co (Publishers) Ltd, London

CIBA-GEIGY
Articles describing applications of epoxy resins have appeared in the following publications:

CIBA Technical Notes, April 1969:	*Araldite in Sydney Opera House*
CIBA Technical Notes, Sept. 1969:	*Araldite in segmental concrete construction*
CIBA-GEIGY Technical Notes, 12/1971:	*Araldite for the grouting of machinery*
CIBA-GEIGY Technical Notes, 4/1972:	*Araldite in a sports stadium (Parc des Princes, Paris)*
CIBA-GEIGY Technical Notes, 8/1972:	*Araldite-based solvent-free coatings*
CIBA-GEIGY Technical Notes, 5/1973:	*Some industrial applications of epoxy resins: grouting-in heavy machine tools*
CIBA-GEIGY Technical Notes, 6/1974:	*Araldite in the aqueduct of Segovia*
CIBA-GEIGY Technical Notes, 10/1974:	*Araldite helps to position Graythorpe I*
CIBA-GEIGY Technical Notes, 12/1974:	*Araldite mortar secures South America's longest bridge*
CIBA-GEIGY Technical Notes, 3/1976:	*Araldite bonds the Olympic Stadium (Montreal)*
CIBA-GEIGY Technical Notes, 3/1978:	*Araldite in crane rail installations*
CIBA-GEIGY Resin Aspects, No. 22:	*Polymer concrete machine bases offer enhanced performance*
CIBA POLYMERS Aspects, No. 34:	*World's longest composite bridge bonded with Araldite*

CLARKE, JL (editor) (1993)
Alternative Materials for the Reinforcement and Prestressing of Concrete
Blackie Academic and Professional

COMITE EURO-INTERNATIONAL DU BETON (1994)
Fastenings to Concrete and Masonry Structures: State of the Art Report
Thomas Telford, London

CONSTRUCTION INDUSTRY RESEARCH AND INFORMATION ASSOCIATION (1987)
Protection of reinforced concrete by surface treatments (Technical Note 130)
CIRIA/Butterworths, London

COWIE, JMG (1991)
Polymers: chemistry and physics of modern materials
Blackie, Glasgow and London

DHIR, RK and GREEN, JW (editors) (1990)
Protection of Concrete – Proceedings of the International Conference, University of Dundee, 11–13 September 1990
E and FN Spon, London

DORAN, DK (editor) (1992)
Construction materials reference book
Butterworth–Heinemann Ltd, Oxford

EI-BADRY, MM (editor) (1996)
Advanced composite materials in bridges and structures: 2nd International Conference, Montreal 11–14 August
The Canadian Society for Civil Engineering

FELDMAN, D (1989)
Polymeric building materials
Elsevier Applied Science, London and New York

FeRFA(1979)
The use of resins in concrete construction and repair
Seminar held in London, 18 October 1979

FeRFA(1980)
Resins for industrial problem areas: flooring and anti-corrosion applications
Seminar held in London, 11 November 1980

FeRFA(1984)
Resins in construction – 20+ years experience
Seminar held in London, 25 October 1984

FeRFA (1989)
Reinforced concrete – structural considerations and choice of products for repairs
(Technical Report)

FeRFA (1991)
Resin mortar bedding for bridge bearing pads
Application Guide No. I

FIP COMMISSION ON PRACTICAL CONSTRUCTION (1991)
Repair and strengthening of concrete structures – FIP guide to good practice
Thomas Telford, London

GALKA, R and YATES, APJ (1984)
Pipe protection – a review of current practice
The British Hydromechanics Research Association (with the Paint Research Association) Bedford

HAIGH, IP (editor) (1982)
Painting steelwork (Report 93)
Construction Industry Research and Information Association, London

HARE, CH (1994)
Protective coatings – fundamentals of chemistry and composition
Technology Publishing Company, Pennsylvania

HEWLETT, PC and SHAW, JDN (1977)
Structural adhesives in civil engineering
In: Developments in Adhesives I (WC Wake editor), Chapter 2, pp 25–75
Applied Science Publishers, New York

HOLLOWAY, L (editor) (1990)
Polymers and polymer composites in construction
Thomas Telford, London

HOLLOWAY, L and LEEMING, MB (editors) (1999)
Strengthening of reinforced concrete structures
Woodhead Publishing Ltd, Cambridge

HUTCHINSON, A R (1997)
Joining of fibre-reinforced polymer composite materials (Project Report 46)
Construction Industry Research and Information Association, London

INTERNATIONAL CONGRESS ON POLYMERS IN CONCRETE
Polymers in Concrete, London 5–7 May 1975
The Construction Press (for the Concrete Society)

INTERNATIONAL CONGRESS ON POLYMERS IN CONCRETE
Polymers in Concrete, Austin Texas, 25–27 October 1978
The University of Austin, USA

INTERNATIONAL CONGRESS ON POLYMERS IN CONCRETE
Polymers in Concrete, Koriyama, Japan, 13–15 May 1981 (2 Vols)
Nihon University, Japan

INTERNATIONAL CONGRESS ON POLYMERS IN CONCRETE
Polymers in Concrete, Darmstadt, Germany, 19–21 September 1984
Institut für Spanende Technologie und Werkzeugmaschinen, Darmstadt

INTERNATIONAL CONGRESS ON POLYMERS IN CONCRETE
The Production, Performance and Potential of Polymers in Concrete, Brighton, England, 22–24 September 1987
Brighton Polytechnic, Sussex, England

INTERNATIONAL CONGRESS ON POLYMERS IN CONCRETE
Polymers in Concrete, Shanghai, 24–27 September 1990
International Academic Publishers

INTERNATIONAL CONGRESS ON POLYMERS IN CONCRETE
Polymers in Concrete, North American Workshop, San Francisco, 24–27 September 1991

INTERNATIONAL CONGRESS ON POLYMERS IN CONCRETE
Polymers in Concrete, Moscow, 22–25 September 1992
Science Business Contact International

INTERNATIONAL CONGRESS ON POLYMERS IN CONCRETE
Polymers in Concrete, Ostende, Belgium, 3–5 July 1995
Royal Flemish Society of Engineers

INTERNATIONAL CONGRESS ON POLYMERS IN CONCRETE
Polymers in Concrete, Bologna, Italy, 14–18 September 1998
Italian Association of Materials Engineers (AIMAT)

INTERNATIONAL JOURNAL OF ADHESION and ADHESIVES (1982)
Adhesives in Civil Engineering: Special Issue, Vol. 2, No. 2, April 1982,
Butterworth Scientific Limited, Surrey

KLAIBER, FW, DUNKER, KF, WIPF, TJ AND SAUNDERS, WW (1987)
Methods of strengthening existing highway bridges
National Cooperative Highway Research Programme, Report 293,
Transportation Research Board, Washington DC

KLOSOWSKI, JM (1989)
Sealants in construction
Marcel Dekker Inc, New York

MAILVAGANAM, NP (editor) (1992)
Repair and protection of concrete structures
CRC Press, Florida

MALLETT, GP (1994)
Repair of concrete bridges: state of the art review
Thomas Telford, London

MAYS, GC (editor) (1992)
Durability of concrete structures
E and FN Spon, London

MAYS, GC and HUTCHINSON, AR (1992)
Adhesives in civil engineering
Cambridge University Press, Cambridge

MIKHAILOV, KV, PATUROEV, VV and KREIS, R (1992)
Polymer concretes and their structural uses
AA Balkenna Publishers, Rotterdam

NANNI, A and DOLAN, CW (editors) (1993)
Fibre-reinforced – plastic reinforcement for concrete structures, International Symposium, Vancouver 28–31 March (ACI SP-138)
The American Concrete Institute, Detroit

OHAMA, Y (1984)
Polymer-modified mortars and concretes
Concrete Admixtures Handbook: Properties, Science and Technology
(VS Ramachandran, editor)
Noyes Publications, New Jersey

OHAMA, Y (1987 and 1990)
Bibliography on polymers in concrete
International Congress on Polymer in Concrete

PANEK, JR and COOK, JP (1984)
Construction sealants and adhesives
John Wiley and Sons, New York

PATERSON, WS (1973)
Fixings on concrete and masonry: an appraisal of present knowledge
Construction Industry Research and Information Association (Technical Note 51), London

PATERSON, WS (1977)
Selection and use of fixings in concrete and masonry construction
Construction Industry Research and Information Association (Guide 4), London

PERKINS, PH (1986)
Repair protection and waterproofing of concrete structures
Elsevier Applied Science Publishers Ltd, England

POTTER, WG (1975)
Uses of epoxy resins
Newnes-Butterworths, London

PULLAR-STRECKER, P (1987)
Corrosion damaged concrete – assessment and repair
CIRIA/Butterworths, London

REID, IW (1986)
GRP sanitary sewer pipes – integral corrosion resistant linings
Construction Industry Research and Information Association, London, Report 112

RILEM (1967)
Synthetic resins in building construction
Symposium held in Paris, 4–6 September 1967

RILEM (1995)
Disposal and recycling of organic and polymeric construction materials
Tokyo Seminar (Y Ohama, editor)

SASSE, HR (editor) (1986)
Adhesion between Polymers and concrete: Proceedings of an International Symposium organised by RILEM Technical Committee 52, Aix-en-Provence, 16–19 September 1986
Chapman and Hall, London/New York

SAUNDERS, KJ (1973)
Organic polymer chemistry
Chapman and Hall, London

SEYMOUR, RB (editor) (1979)
Plastic mortars, sealants and caulking compounds
(American Chemical Society Symposium Series 13)
American Chemical Society, Washington DC

SHAW, JDN (1990)
Adhesives in the construction industry: materials and case histories
Construction and Building Materials, Vol. 4, No.2, pp.92–97, June 1990

TABOR, LJ (1978)
Effective Use of Epoxy and Polyester Resins in Civil Engineering
Construction Industry Research and Information Association, London, Report 69

TAERWE, L (editor) (1995)
Non-metallic (FRP) reinforcement for concrete structures; Second International RILEM Synposium, (FRPRCS-2) 23–25 August, Ghent, Belgium
E + FN Spon, London

THE CONCRETE SOCIETY (1975)
Polymer concrete (Concrete Society Technical Report No 9)
The Concrete Society, Crowthorne

THE CONCRETE SOCIETY (1984)
Repair of concrete damaged by reinforcement corrosion
Concrete Society (with FeRFA and the Association of Gunite Contractors)
Technical Report 26, Crowthorne

THE CONCRETE SOCIETY (1994)
Polymers in concrete
Concrete Society Technical Report 39, Crowthorne

THE PLASTICS INSTITUTE/INSTITUTION OF CIVIL ENGINEERS (1973)
Resins and Concrete Symposium, University of Newcastle-upon-Tyne,
17–18 April 1973

TRANSPORT AND ROAD RESEARCH LABORATORY (1974)
Proceedings of the Symposium on Skid-Resistant Roads
Crowthorne (TRRL Supplementary Report 24 UC)

TUTT J N (1988)
Replacement ties in cavity walls
Construction Industry Research and Information Association (Report 117), London

UNIVERSITY OF SHEFFIELD (1990)
International Seminar on Structural Repairs/Strengthening by the Plate Bonding Technique

WRIGHT N (1978)
Recent developments in surface dressing in the UK
Transport and Road Research Laboratory, Crowthorne
(TRRL Supplementary Report 486)